Smart and Sustainable Cities?

Smart cities promise to generate economic, social and environmental value through the seamless connection of urban services and infrastructure by digital technologies. However, there is scant evidence of how these activities can enhance social well-being and contribute to just and equitable communities.

Smart and Sustainable Cities? Pipedreams, Practicalities and Possibilities provides one of the first examinations of how smart cities relate to environmental and social issues. It addresses the gap between the ambitious visions of smart cities and the actual practices on the ground by focusing on the social and environmental dimensions of real smart city initiatives as well as the possibilities they hold for creating more equitable and progressive cities. Through detailed analyses of case studies in the United States, Australia, the United Kingdom, Japan, Germany, India and China, the contributors describe the various ways that social and environmental issues are interpreted and integrated into smart city initiatives and actions. The findings point towards the need for more intentional engagement and collaboration with all urban stakeholders in the design, development and maintenance of smart cities to ensure that everyone benefits from the increasingly digitalised urban environments of the twenty-first century.

The chapters in this book were originally published as a special issue of the journal *Local Environment*.

James Evans is Professor of Geography at the University of Manchester, UK.

Andrew Karvonen is Associate Professor in the Division of Urban and Regional Studies at the KTH Royal Institute of Technology, Sweden.

Andres Luque-Ayala is Associate Professor of Geography at Durham University, UK.

Chris Martin was formerly Research Associate in the Department of Geography at the University of Manchester, UK.

Kes McCormick is Associate Professor at the International Institute for Industrial Environmental Economics (IIIEE), Lund University, Sweden.

Rob Raven is Professor at the Monash Sustainable Development Institute, Monash University, Australia.

Yuliya Voytenko Palgan is Associate Professor at the International Institute for Industrial Environmental Economics (IIIEE), Lund University, Sweden.

Smart and Sustainable Cities?

Pipedreams, Practicalities and Possibilities

Edited by
**James Evans, Andrew Karvonen,
Andres Luque-Ayala, Chris Martin,
Kes McCormick, Rob Raven and
Yuliya Voytenko Palgan**

LONDON AND NEW YORK

First published 2021
by Routledge
2 Park Square, Milton Park, Abingdon, Oxon OX14 4RN

and by Routledge
52 Vanderbilt Avenue, New York, NY 10017

Routledge is an imprint of the Taylor & Francis Group, an informa business

British Library Cataloguing in Publication Data
A catalogue record for this book is available from the British Library

ISBN 13: 978-0-367-63680-7

Typeset in Myriad
by Newgen Publishing UK

Publisher's Note
The publisher accepts responsibility for any inconsistencies that may have arisen during the conversion of this book from journal articles to book chapters, namely the inclusion of journal terminology.

Disclaimer
Every effort has been made to contact copyright holders for their permission to reprint material in this book. The publishers would be grateful to hear from any copyright holder who is not here acknowledged and will undertake to rectify any errors or omissions in future editions of this book.

Contents

Citation Information

The chapters in this book were originally published in *Local Environment*, volume 24, issue 7 (September 2019). When citing this material, please use the original page numbering for each article, as follows:

Chapter 6

Smart and eco-cities in India and China
Johanna I. Höffken and Agnes Limmer
Local Environment, volume 24, issue 7 (September 2019), pp. 646–661

For any permission-related enquiries please visit:
www.tandfonline.com/page/help/permissions

Notes on Contributors

Jess Britton, Energy Policy Group (EPG), University of Exeter, UK.

James Evans, University of Manchester, UK.

Johanna I. Höffken, Eindhoven University of Technology, Netherlands.

Andrew Karvonen, KTH Royal Institute of Technology, Stockholm, Sweden.

Matthias Lehner, International Institute for Industrial Environmental Economics (IIIEE), Lund University, Sweden.

Anthony M. Levenda, School for the Future of Innovation in Society, Arizona State University, Tempe, AZ, USA.

Agnes Limmer, Technical University of Munich, Germany.

Heather Lovell, School of Social Sciences, University of Tasmania, Hobart, Australia.

Andres Luque-Ayala, Durham University, UK.

Chris Martin, formerly at the University of Manchester, UK.

Kes McCormick, Lund University, Sweden.

Oksana Mont, International Institute for Industrial Environmental Economics (IIIEE), Lund University, Sweden.

Andrius Plepys, International Institute for Industrial Environmental Economics (IIIEE), Lund University, Sweden.

Rob Raven, Monash University, Melbourne, Australia.

Gregory Trencher, Graduate School of Environmental Studies, Tohoku University, Sendai, Japan.

Yuliya Voytenko Palgan, International Institute for Industrial Environmental Economics (IIIEE), Lund University, Sweden.

Lucie Zvolska, International Institute for Industrial Environmental Economics (IIIEE), Lund University, Sweden.

Introduction

Smart and sustainable cities? Pipedreams, practicalities and possibilities

1. Introduction

Smart cities promise to generate economic, social and environmental value through the seamless connection of urban services and infrastructure by digital technologies (Hollands 2008; Viitanen and Kingston 2014), but there is scant evidence concerning their ability to enhance social well-being, build just and equitable communities, reduce resource consumption and waste generation, improve environmental quality or lower carbon emissions (Cavada, Hunt, and Rogers 2016). This special issue addresses the gap between the pipedream and the practice of smart cities, focusing on the social and environmental dimensions of real smart city initiatives, and the possibilities that they hold for creating more equitable and progressive cities. We argue that social equity and environmental sustainability are neither a-priori absent nor de-facto present in technological designs of smart city initiatives, but have to be made, nurtured and maintained as they materialise in particular places. This is the 'possibility' alluded to in our title, and where the focus of the Special Issue on the gap between the pipedreams and practicalities of smart cities leads. In this introduction we unpack our argument in greater detail and situate our six contributions within it.

Smart cities promise nothing less than an urban utopia for the twenty-first century (Datta 2015). Stoked by estimates of a global smart city market worth up to $1.56 trillion by 2020 (Frost & Sullivan 2014), the concept has risen rapidly to prominence within industry, political and municipal discourses of urban development (Söderström, Paasche, and Klauser 2014). In 2015, de Jong and colleagues predicted that the smart city was 'on its way to become [the] leading driver of urban sustainability and regeneration initiatives' (12). In the European Union, digital innovations now underpin the majority of urban sustainability funding, with the creation of smart cities commonly positioned as a vehicle to deliver urban sustainability and economic growth (European Parliament 2014). Across Africa, South Asia, North America and the Far East, national governments, municipalities and private companies are allocating considerable resources to develop digital innovations which they claim will promote a vibrant entrepreneurship culture in cities, advance more equitable and just community development through increased citizen participation, and solve a range of sustainability issues such as climate change.

The smart city pipedream diverges from other urban utopias in three quite distinctive ways. First, smart cities occupy mainstream policy and thinking, unlike the utopian settlements of the nineteenth century that were by definition counter cultural and limited to progressive colonial movements and the model villages of industrial philanthropists. Second, the smart city utopia reflects a close union between national government and private industry. The Corbuserian dream of towers in the countryside that inspired much post-war building in both the East and West in the twentieth century was driven by the government rather than industry. These government-led programmes accompanied the emergence of the welfare state while today's smart city visions are based firmly on entrepreneurialism and profit-seeking public-private partnerships. Finally, while previous urban utopias were inspired by explicit political and social goals, smart cities foreground economic development as the main driver to realise future cities (Haarstad 2017).

These differences are significant when we start to think about the power of digital technologies to make cities more sustainable. The idea of smart and sustainable cities promises to use digital technologies to make infrastructure services more efficient and reactive to user behaviour, lower resource consumption, increase environmental quality, and cut down on carbon emissions. It is this alleged convergence of 'smartness' with urban sustainability that provides the starting point for this special issue. While the rise of the smart city approach places digital innovation, the digital economy and urban growth at the centre of efforts to create sustainable cities (Viitanen and Kingston 2014), the ability of smart technology to deliver social and environmental sustainability remains little more than an article of faith.

One need not look far to find a series of underlying tensions between the logics of the smart city and sustainable urban development (Martin, Evans, and Karvonen 2018; Marvin et al. 2019). Some contradictions are so great they have inspired active resistance against the smart city dream (Greenfield 2013). For example, smart urban development tends to reinforce neoliberal economic growth and consumerist culture (Vanolo 2014), focusing on more affluent populations who are able to access private services like Uber and Airbnb. Smartness reframes urban sustainability challenges as market opportunities for corporations to sell digital solutions (Viitanen and Kingston 2014) with has implications for how we conceptualise urban sustainability. For example, the challenge of providing clean energy to large urban populations becomes a question of providing smart meters, deploying smart grids and managing them using some form of digital urban operating system rather than developing new forms of community energy provision or use (see Britton 2019). The challenge of managing urban waste is reframed as a matter of optimising urban logistical flows through algorithmic calculation (i.e. the optimisation of waste collection routes) rather than considering issues of resource consumption. These tendencies mean that smart cities risk marginalising citizens, prioritising end-of-pipe solutions, and driving further economic development that runs counter to stated environmental or social objectives.

This special issue brings together an interdisciplinary collection of articles to present a detailed empirical analysis of how smart city approaches are reshaping the social and environmental dimensions of sustainable urban development on the ground. There is a shared focus on how the contradictions between sustainability and digital urbanism are being negotiated in practice. The motivations for the special issue are twofold. First, the issue responds to a genuine trend within both government and corporate smart city discourse towards collapsing smart and sustainability narratives, a move that is often heralded through the inclusion of citizens and communities as key stakeholders within the making of 'smart' plans and initiatives. The second motivation is a response to the lack of detailed studies on how smart initiatives are unfolding on the ground, particularly in relation to issues of social and environmental sustainability. Rather than remaining fixated on the endless iterations of technological triumphalism found in strategy documents and corporate brochures, there is a growing need to understand how the smart city discourse is actually landing in and transforming ordinary cities and communities (Luque-Ayala and Marvin 2015; Shelton, Zook, and Wiig 2015; Wiig and Wyly 2016; Karvonen, Cugurullo, and Caprotti 2019).

Of particular interest is how smart cities are influencing social issues of equity, justice, citizen participation, health and wellbeing. Identifying and assessing the deficits and potentialities of different forms of smart and sustainable urban development to address social and environmental challenges is the first step in providing an evidence base for alternative and potentially more progressive approaches to smart city development at local levels (Glasmeier and Christopherson 2015; Glasmeier and Nebiolo 2016). For example, digital technologies present significant potential to improve access to basic services like water and fuel in informal settlements – a move that would free up time for education and work, enhance safety and health and enrol residents in collaborative modes of governance. Similar arguments can be made concerning the potential of technology to improve social care and facilitate sharing schemes in developed world contexts. The challenge here is that while social impacts are considerable, direct economic benefits of the kind value by government decision-makers are more difficult to capture.

Despite the relative newness of the smart city concept, the contributions of the special issue both acknowledge and interrogate the extent to which conceptualizations and practices of smart city development have changed over time – and continue to do so. An initial technological push version of smart city development (c.f. Luque-Ayala and Marvin 2016; McNeill 2015), characterised as a form of 'solutionism' (Morozov 2013) whereby private companies offer municipalities large scale digital solutions, often looking for urban problems to solve, failed to take hold. Meanwhile cash-strapped municipalities, hesitant to part with large amounts of cash for systems that they are neither sure they need nor keen to become locked into, have started to seek alternative mechanisms to procure, adopt and promote digital technologies. An emerging ecosystem of urban digital technologies, often made up of small start-ups and experimental projects, transcended the market space initially created by big players such as IBM, Cisco or Hitachi.

A more recent incarnation of the smart city discourse flips the technological push version of the smart city on its head to start from the needs and requirements of citizens, preferably with citizen involvement (Trencher forthcoming). This version of the smart city aims to improve living standards but requires social awareness from citizens to work (Bhagya et al. 2018). A range of tools have emerged to promote citizen involvement, including data platforms and urban living labs, which are intended to empower the public to engage with processes of urban governance via digital technologies (Voytenko et al. 2016). The 'Smart with a Heart' tagline from the 2018 Nordic Edge Smart City Expo captures this trend toward the people-centred smart city perfectly. Yet, the catch with citizen involvement in practice is that the major players (industry, local government) often lack the resources, time, skills or aspirations to engage people effectively, or only do so in the later stages of initiatives when citizens have limited power to shape change (Menny, Voytenko Palgan, and McCormick 2018). Smart city programmes that have been genuinely steered through engagement with residents, or advanced alternative or novel notions of urban and economic development, social and political inclusion, or greater environmental stewardship are thin on the ground (Martin, Evans, and Karvonen 2018).

2. Contributions to the special issue

Thrust by public and corporate interests into the front and centre of current urban practice, smart cities are at a cross roads. While enjoying something of a 'moment' it remains unclear whether they will be a force either for or against sustainability. This special issue emerges from paper sessions at three academic conferences in 2016: the American Association of Geographers (AAG) Annual Meeting in San Francisco, the Royal Geographical Society with the Institute of British Geographers (RGS-IBG) Annual International Conference in London, and the joint meeting of the Society for the Social Study of Science (4S) and the European Association for the Study of Science and Technology (EASST) in Barcelona. The aim of these sessions was to go beyond the corporate and policy hype that surrounds smart cities to understand what is actually happening on the ground. It reflects an increasing interest in understanding the social and environmental dimensions of smart city developments that engage national and municipal policies and politics in shaping the agenda over time and space, and the role of different forms of smart technology in producing new kinds of social, economic and political relations.

At the outset, we asked what role visions, discourses and practices of smart design and thinking are playing in changing how we imagine a sustainable city will look like, how it will function, and who it will serve. It is commonplace to find smart demonstration projects, districts and experiments in cities around the globe (Voytenko et al. 2016). Increasingly smart sustainable urbanism is synonymous with local, project-driven actions that are intended to demonstrate how smart can deliver sustainability and, on the basis of this, promote the subsequent rolling out, replicating or upscaling of solutions (Evans, Karvonen, and Raven 2016). We ask how these phenomena are reshaping cities and what their implications for urban sustainability are. Which groups of people and organisations are empowered or disempowered by smart approaches to sustainability and with what effects? What types of interests, values, competencies and evidence are being privileged and neglected by smart sustainability? Finally, we are interested in what all this amounts to for cities and their residents.

Is there a distinct form of smart sustainable governance emerging and if so what kind of city is it creating?

The six articles in this special issue reveal key dimensions of urban smart-sustainability and their social and environmental implications from a range of different perspectives and in a range of different contexts. The contributions are international, comprising cases from Germany, Australia, the UK, the USA, Japan, India and China. They are also interdisciplinary, with authors drawing on debates from the fields of Urban Studies, Geography, Environmental Sciences, Political Science, Planning, and Science and Technology Studies. The first three articles focus on local demonstrations of smart technology that are intended to enhance sustainability, and reflect on their social and environmental aspects. The fourth and fifth articles explore ways in which smart-sustainability is being stretched in specific cities, to address public health and resource sharing respectively. The final article zooms out to examine how smart-sustainability is shaping new forms of governance at the national scale in India and China.

In our first contribution, Anthony Levenda investigates a smart grid experimentation and demonstration site in a residential neighbourhood in Austin, Texas. The analysis shows how this urban living lab is a product of Austin's 'technopolis', and interprets this smart grid experimentation as a form of governmentality that devolves urban imperatives into individual responsibilities for socio-environmental change. The management of carbon and electricity use through energy efficiency, renewable energy, and conservation is promoted as a form of self-management, wherein households must reconfigure everyday practices and/or adopt new technologies. At the same time, the majority of these interventions are created in a top-down fashion, shaped by technology companies, researchers, and policy-makers. This skews the potential of active co-production, and instead relies on the delegation of responsibility for action to a limited assemblage of smart technologies and smart users.

Heather Lovell picks up this theme through her examination of smart grids in Australia, which highlights both the potential and pitfalls of digitally-enabled green urbanism. Empirical research on two Australian smart grid initiatives – the Smart Grid Smart City project and the State of Victoria's Advanced Metering Infrastructure Programme – provides key insights into how and why the high-tech data-led intelligence of smart grids has encountered problems at the point of implementation. Conceptually, the article draws on the concept of governmentality to show how the promise of smart grids has not been realised in practice, in large part because new digital technologies have not 'behaved' in the way originally planned. This undermined assumptions that the smart technologies would work to support government programmes and created a failure of governance. The paper shows how greater understanding of and engagement with the people receiving and using these technologies is required in order to realise the societal benefits of smart cities.

In the third article, Jess Britton examines the installation of domestic gas and electricity smart meters across the UK. This state-led initiative is providing an unprecedented volume and granularity of energy consumption data with the aim of achieving ambitious energy efficiency and carbon emission reduction goals. Smart meter programmes typically target individual energy consumers while providing network-wide opportunities for commercial applications. Britton argues that in addition to these two scales, the city scale has the potential to unlock public interest benefits through applications to public policymaking, community energy projects, and fuel poverty campaigns. However, the current arrangements for the access and use of smart meter data present a number of challenges related to complexity, dominance of incumbent actors, data access and uncertainty that become barriers to city-scale actions. There is a need to overcome these barriers to realise the collective benefits of smart-sustainable cities for all of society rather than a select few.

A perennially overlooked objective in both smart and sustainable urban development projects is that of human experience and well-being. Can smart, sustainable cities move beyond narrow ambitions of technological innovation, economic prosperity and reducing environmental impacts to actually facilitate healthier, happier and more fulfilling lifestyles? If so how? And what factors might stand in the way of such a 'stretched' smart city agenda? In their article, Gregory Trencher and Andrew Karvonen explore how health and lifestyle goals are being pursued in a Japanese smart city. The

Kashiwa-no-ha Smart City in greater Tokyo uses ICT technologies to provide preventative health services for elderly residents whilst implementing various interventions to promote active, socially rich and environmentally aware lifestyles. The analysis reveals how smart technologies can be used to address widespread and relevant problems that are relevant to specific groups of urban residents.

A second important trend in terms of social and environmental applications involves the role of smart cities in facilitating the emergence and diffusion of sharing (McLaren and Agyeman 2015). European cities, being attuned to the values of fairness, sustainability, and cooperation, are envisaged to catalyse the sharing economy and circumvent its corporatisation. However, there is a lack of knowledge on how cities are bringing smart, sustainable and sharing agendas together. Lucie Zvolska, Matthias Lehner, Yuliya Voytenko Palgan, Oksana Mont and Andrius Plepys explore the role of city governments in advancing sustainability via ICT-enabled sharing, focusing on the case studies of Berlin and London – two ICT-dense cities with clearly articulated smart city agendas and an abundance of sharing platforms. The article provides insights into how city governments are supporting ICT-enabled sharing platforms, discussing the ways in which these platforms both advance and stand in the way of urban sustainability.

The final article broadens the perspective to look at how smart sustainability varies across different national contexts and how this influences the governance of cities. Johanna Höffken and Agnes Kneitz give us a glimpse into the current political landscape of India and China, where technological visions of the sustainable city are currently being turned into political and concrete reality through India's Smart City Mission and China's Ecological Civilization. Though there are clear differences in their respective approach to the mutual problem of population growth and social change related to urbanisation, both approaches share interesting similarities and allow for a rich transnational comparison. In India under the leadership of Prime Minister Narendra Modi, 100 cities are to be transformed into smart habitats. Meanwhile, Chinese president Xi Jinping modestly aims at transforming its entire society into a green life form, and envisions becoming a global beacon for ecological civilisation and citizenship. This article contributes important non-Western perspectives on the politics of participation in the smart-sustainable city. The authors ask what groups of actors participate in negotiating and realising the eco-smart approaches to urban sustainability, and analyses the way they are doing so. Why are they invested in their national projects of greening themselves in a potentially smart way? How does that reflect back into policy making and is there potential to learn from each other?

3. Conclusions: Equity and environment in the smart-sustainable city

The twenty-first century has been hailed as the urban century, and one in which 'smartness' will shape urban responses to global challenges (McCormick et al. 2013). In the past few years scholars have been trying to understand 'why, how, for whom and with what consequences' the smart city paradigm emerges in different urban contexts (Luque-Ayala and Marvin 2015, 2106). This special issue focuses on how the smart agenda is being combined with notions of sustainable urban development. Both smart and sustainability narratives have been widely critiqued for promoting economic development while giving scant attention to environmental and social concerns. No doubt it is commonplace for proponents of the smart-sustainable city to put forward a narrow and technocratic perspective that reduces sustainability to a series of 'technical and economic fixes' (Bulkeley and Betsill 2005, 58) while obscuring its political implications (Gibbs and Krueger 2007).

At the same time, there are synergies in smart-sustainable cities that either go beyond or could potentially go beyond broadly neoliberal 'business as usual'. Making urban data widely available, developing a city-scale 'Internet of Things' and mobilising smart and digital technologies has the potential to enhance social well-being, empower communities, reveal previously hidden urban environmental processes, enable resource and skills sharing, include citizens in co-creative governance processes, generate novel solutions to mundane urban problems, and underpin new models for more efficient use of idle assets (see Zvolska et al. 2019; Menny, Voytenko Palgan, and McCormick

2018; McLaren and Agyeman 2015; Martin, Evans, and Karvonen 2018). But this also requires the careful application of ICT technologies to avoid empowering corporate interests within urban visioning and management and further excluding those already marginalised by prevailing technocratic and entrepreneurial forms of urban governance (Vanolo 2014; Söderström, Paasche, and Klauser 2014; Hollands 2016). The varieties of smart-sustainable agendas emerging on the ground in hundreds of cities around the world occupy a wide range of positions between (and beyond) these critiques and promises. Social equity and environmental sustainability are not a-priori absent or de-facto present in technological designs of smart city initiatives, but have to be made, nurtured and maintained, as they materialise in particular places.

As a whole, the articles in this collection represent an emerging agenda that has the potential to open up smart-sustainable urban development to a broader coalition of actors and achievements. But for this to happen, city governments, communities, tech-start-ups, corporates and knowledge institutes need to steer smart urban development to address issues that are relevant to their particular contexts and concerns. Smart-sustainable applications need to be aligned with neighbourhood and city scales rather than focus only on individual consumer behaviour and international commercial agendas. If smart is to enable sustainable urban development, this suggests a scalar politics of smart-sustainable cities in which collective agendas and visions have to be constructed around particular issues of social and environmental relevance at the local and city level. Contextual factors such as historical development patterns, cultural norms and practices, and political structures have a significant influence on how smart-sustainable is rolling out and generating momentum in particular places (Raven et al. 2017). In this sense, 'the relationship between smart technologies and urban environments is therefore recursive, with each serving to transform the other' (Kong and Woods 2018, 685).

One of the most significant potential implications of the smart-sustainable city is its implications for the knowledge politics of urban development (de Hoop et al. 2018; Cowley and Caprotti forthcoming). The articles in this special issue demonstrate how an economically informed (e.g. austerity-led) pursuit of innovation is disrupting traditional modes of governance in positive and negative ways, blindly reifying notions of efficiency and optimisation – but also foregrounding the benefits of demonstration, testing, and co-production. This has implications for how cities are steered and by whom. Sensors, digital infrastructures, machine learning, urban dashboards, digital platforms and smart phone apps are changing the ways we see and know our cities and, if acted upon, may have far-reaching implications for how urban change is directed. Rather than simply suggesting a neoliberal 'takeover' of urban governance occurring through the digitalisation of networked infrastructures, the emerging wave of smart urbanism potentially points towards a post-networked form of urban governance (c.f. Coutard and Rutherford 2016), activated via material, social and political forms of decentralisation operating in tandem and with implications that may both narrow as well as open up social and environmental urban sustainability. Ultimately, we as academics must engage with the often messy and frustrating processes of urban development and strategy in order to steer smart city agendas and actions in more progressive directions. If we do not ensure that social and environmental interests have a seat at this table, we will surely find them on the menu.

Funding

This work was supported by the Svenska Forskningsrådet Formas [grant number 2014-6366-27310-35].

References

Bhagya, N., M. Silva, M. Khan, and H. Kijun. 2018. "Towards Sustainable Smart Cities: A Review of Trends, Architectures, Components, and Open Challenges in Smart Cities." *Sustainable Cities and Society* 38: 697–713.

Britton, J. 2019. "Smart Meter Data and Equitable Energy Transitions – Can Cities Play a Role?" *Local Environment*. doi:10.1080/13549839.2017.1383372.

Bulkeley, H., and M. Betsill. 2005. "Rethinking Sustainable Cities: Multilevel Governance and the 'Urban' Politics of Climate Change." *Environmental Politics* 14 (1): 42–63.

Cavada, M., D. V. Hunt, and C. D. Rogers. 2016. "Do Smart Cities Realise their Potential for Lower Carbon Dioxide Emissions?" *Proceedings of the Institution of Civil Engineers-Engineering Sustainability* 169 (6): 243–252.

Coutard, O., and J. Rutherford, eds. 2016. *Beyond the Networked City: Infrastructure Reconfigurations and Urban Change in the North and South.* London: Routledge.

Cowley, R., and F. Caprotti. forthcoming. "Smart City as Anti-planning in the UK." *Environment and Planning D: Society and Space.*

Datta, A. 2015. "New Urban Utopias of Postcolonial India: 'Entrepreneurial Urbanization' in Dholera Smart City, Gujarat." *Dialogues in Human Geography* 5 (1): 3–22.

de Hoop, E., A. Smith, W. Boon, R. Macrorie, S. Marvin, and R. Raven. 2018. "Smart Urbanism in Barcelona: A Knowledge Politics Perspective." In *The Politics of Urban Sustainability Transitions: Knowledge, Power and Governance,* edited by J. Jensen, P. Spath, and M. Cashmore, 33–52. Routledge.

European Parliament. 2014. *Mapping Smart Cities in the EU.* Brussels: European Union.

Evans, J., A. Karvonen, and R. Raven, eds. 2016. *The Experimental City.* London: Routledge.

Frost & Sullivan. 2014. Global Smart Cities Market to Reach US$1.56 Trillion by 2020 [WWW Document]. Accessed January 18, 2016. http://ww2.frost.com/news/press-releases/frost-sullivan-global-smart-cities-market-reach-us156-trillion-2020.

Gibbs, D., and R. Krueger. 2007. "Containing the Contradictions of Rapid Development? New Economy Spaces and Sustainable Urban Development." In *The Sustainable Development Paradox: Urban Political Economy in the United States and Europe,* edited by R. Krueger, and D. Gibbs, 95–122. New York: The Guilford Press.

Glasmeier, A., and S. Christopherson. 2015. "Thinking About Smart Cities." *Cambridge Journal of Regions, Economy and Society* 8: 3–12.

Glasmeier, A. K., and M. Nebiolo. 2016. "Thinking About Smart Cities: The Travels of a Policy Idea that Promises a Great Deal, but So Far has Delivered Modest Results." *Sustainability* 8: 1122–1133.

Greenfield, A. 2013. *Against the Smart City: A Pamphlet. This is Part I of "The City is Here to Use".* Do projects.

Haarstad, H. 2017. "Constructing the Sustainable City: Examining the Role of Sustainability in the 'Smart City' Discourse." *Journal of Environmental Policy and Planning* 19 (4): 423–437.

Hollands, R. G. 2008. "Will the Real Smart City Please Stand Up?" *City* 12: 303–320.

Hollands, R. G. 2016. "Beyond the Corporate Smart City? Glipses of Other Possibilities of Smartness." In *Smart Urbanism: Utopian Vision or False Dawn?,* edited by S. Marvin, A. Luque-Ayala, and C. McFarlane, 168–183. London: Routledge.

Karvonen, A., F. Cugurullo, and F. Caprotti, eds. 2019. *Inside Smart Cities: Place, Politics and Urban Innovation.* London: Routledge.

Kong, L., and O. Woods. 2018. "The Ideological Alignment of Smart Urbanism in Singapore: Critical Reflections on a Political Paradox." *Urban Studies* 55 (4): 679–701.

Luque-Ayala, A., and S. Marvin. 2015. "Developing a Critical Understanding of Smart Urbanism?" *Urban Studies* 52 (12): 2105–2116.

Luque-Ayala, A., and S. Marvin. 2016. "The Maintenance of Urban Circulation: An Operational Logic of Infrastructural Control." *Environment and Planning D: Society and Space* 34 (2): 191–208.

Martin, C. J. 2016. "The Sharing Economy: A Pathway to Sustainability or a Nightmarish Form of Neoliberal Capitalism?" *Ecological Economics* 121: 149–159.

Martin, C. J., J. Evans, and A. Karvonen. 2018. "Smart and Sustainable? Five Tensions in the Visions and Practices of the Smart-sustainable City in Europe and North America." *Technological Forecasting and Social Change* 133: 269–278.

Marvin, S., H. Bulkeley, L. Mai, K. McCormick, and Y. Voytenko Palgan. 2019. *Urban Living Labs: Experimenting with City Futures.* London: Routledge.

McCormick, K., L. Neij, S. Anderberg, and L. Coenen. 2013. "Advancing Sustainable Urban Transformation." *Journal of Cleaner Production* 50: 1–11.

McLaren, D., and J. Agyeman. 2015. *Sharing Cities: A Case for Truly Smart and Sustainable Cities.* Cambridge, MA: MIT Press.

McNeill, D. 2015. "Global Firms and Smart Technologies: IBM and the Reduction of Cities." *Transactions of the Institute of British Geographers* 40 (4): 562–574.

Menny, M., Y. Voytenko Palgan, and K. McCormick. 2018. "Urban Living Labs and the Role of Users in Co-creation." *GAIA – Ecological Perspectives for Science and Society* 27: 68–77.

Morozov, E. 2013. *To Save Everything, Click Here: Technology, Solutionism, and the Urge to Fix Problems That Don't Exist.* London: Penguin.

Raven, R. P. J. M., F. W. Sengers, P. Spaeth, L. Xie, A. Cheshmehzangi, and M. de Jong. 2017. "Urban Experimentation and Institutional Arrangements." *European Planning Studies,* https://doi.org/10.1080/09654313.2017.1393047.

Shelton, T., M. Zook, and A. Wiig. 2015. "The 'Actually Existing Smart City'." *Cambridge Journal of Regions, Economy and Society* 8 (1): 13–25.

Söderström, O. S., T. Paasche, and F. Klauser. 2014. "Smart Cities as Corporate Story-telling." *City* 18 (3): 307–320.

Trencher, G. forthcoming. "Towards a Smart City 2.0: Smartness as a Tool for Tackling Social Problems." *Technological Forecasting and Social Change.*

Vanolo, A. 2014. "Smartmentality: The Smart City as Disciplinary Strategy." *Urban Studies* 51 (5): 883–898.

Viitanen, J., and R. Kingston. 2014. "Smart Cities and Green Growth: Outsourcing Democratic and Environmental Resilience to the Global Technology Sector." *Environment and Planning A: Economy and Space* 46: 803–819.

Voytenko, Y., K. McCormick, J. Evans, and G. Schwila. 2016. "Urban Living Labs for Sustainability and Low Carbon Cities in Europe: Towards a Research Agenda." *Journal of Cleaner Production* 123: 45–54.

Wiig, A., and E. Wyly. 2016. "Introduction: Thinking Through the Politics of the Smart City." *Urban Geography* 37: 485–493.

Zvolska, L., M. Lehner, Y. Voytenko Palgan, O. Mont, and A. Plepys. 2019. "Urban Sharing in Smart Cities: The Cases of Berlin and London." *Local Environment*. doi:10.1080/13549839.2018.1463978.

James Evans

ⓘ http://orcid.org/0000-0002-2953-1118

Andrew Karvonen

ⓘ http://orcid.org/0000-0002-0688-9547

Andres Luque-Ayala

ⓘ http://orcid.org/0000-0002-8044-3666

Chris Martin

ⓘ http://orcid.org/0000-0003-0474-3939

Kes McCormick

ⓘ http://orcid.org/0000-0002-5093-160X

Rob Raven

ⓘ http://orcid.org/0000-0002-6330-0831

Yuliya Voytenko Palgan

ⓘ http://orcid.org/0000-0001-8342-085X

Thinking critically about smart city experimentation: entrepreneurialism and responsibilization in urban living labs

Anthony M. Levenda ⓘ

ABSTRACT

The proliferation of smart technologies, big data, and analytics is being increasingly used to address urban socio-environmental problems such as climate change mitigation and carbon control. Electricity systems in particular are being reconfigured with smart technologies to help integrate renewable generation, enhance energy efficiency, implement new forms of pricing, increase control and automation, and improve reliability. Many of these interventions are experimental, requiring real-world testing before wider diffusion. This testing often takes place in "urban living labs," integrating urban residents as key actors in experimentation with goals for broader sustainability transitions. In this paper, I investigate one such urban living lab focused on smart grid research and demonstration in a residential neighbourhood in Austin, Texas. I develop a framework based in governmentality studies to critically interrogate urban experimentation. Findings suggest that the focus of experimentation devolves urban imperatives into individual responsibilities for socio-environmental change. Managing carbon emissions through energy efficiency, renewable energy, and conservation is promoted as a form of self-management, wherein households reconfigure everyday activities and/or adopt new technologies. At the same, sociotechnical interventions are shaped by technology companies, researchers, and policy-makers marking a central feature of contemporary urban entrepreneurialism. This skews the potential of active co-production, and instead relies on the delegation of responsibility for action to a constrained assemblage of smart technologies and smart users.

1. Introduction

Cities across the planet have increasingly experimented with various socio-technical and policy interventions in order to respond to calls for sustainability and climate mitigation (Castán Broto and Bulkeley 2013a). New technological interventions ranging from smart homes to urban control rooms have been marketed as solutions to help address these environmental challenges. More broadly, "smart cities" technologies have been positioned as opportunities to improve urban environments and stimulate economic development (Höjer and Wangel 2015; IBM 2015). However, these overlapping urban imperatives do not come without problematics and complications. As numerous scholars have noted, the entrepreneurial, economic-growth agendas of sustainable and smart cities approaches often undercut the ecological promises of urban experiments resulting in a gap between visions and reality (Buck and While 2015; Colding and Barthel 2017; Cugurullo 2017;

Rosol, Béal, and Mössner 2017). A parallel area of critical scholarship has shown how climate change and sustainability interventions, especially facilitated by new smart technologies, tends to reduce questions of responsibility for solving environmental problems down to individual's choices and behaviours (Brand 2007; Braun 2014; Gabrys 2014; Moloney and Strengers 2014; Peattie 2010; Soneryd and Uggla 2015; Vanolo 2013; Wakefield and Braun 2014). This paper contributes to these debates by arguing for closer attention to the connections between these two literatures – entrepreneurialism and responsibilization, respectively – in order to critically study smart and sustainable city experiments. My central argument is three-fold: (1) urban experimentation offers opportunities for entrepreneurial forms of governance, at the scale of the individual and the city, to take hold through smart and sustainable cities agendas; (2) the transformative potential of smart and sustainable cities agendas is undercut by a focus on governing individual consumption instead of systemic change; and (3) these techno-fix agendas often produce inequities that are overshadowed in public discourse by the spectacle of sustainability and smartness.

This argument is supported through a case study of the development of a smart and sustainable neighbourhood in Austin, Texas. I frame this development project in Austin's Mueller neighbourhood as a particular form of urban experimentation, what scholars have called "urban living laboratories" (Bulkeley, Castan Broto, and Maassen 2014; Caprotti and Cowley 2016; Cardullo, Kitchin, and Di Feliciantonio 2017; Evans et al. 2015; Evans, Karvonen, and Raven 2016; Karvonen and van Heur 2014; Voytenko et al. 2016). The concept behind this urban living lab (ULL) was the integration of smart home and smart grid technologies to test their performance (defined variously as efficiency, economic benefits, or lower environmental impacts), and to understand how households utilised these technologies and responded to a variety of experimental interventions.

"Smart grid" is a broad descriptor for information and communication technologies integrated into the electric grid (from generation to transmission and distribution to the end-use consumption). These technologies offer greater control, two-way communication, and more frequent data collection promising a variety of benefits including better grid reliability and resiliency, and potentially, a variety of end-user (or household) benefits (Buchholz and Styczynski 2014). Similarly, smart homes is a descriptor for a variety of household technologies that include remote electronic control and management of smart appliances (ranging from refrigerators to washing machines) that take advantage of real-time data collection and communication (Balta-Ozkan et al. 2013). Much like many cases of urban experimentation (Evans and Karvonen 2014; Moloney and Horne 2015; Voytenko et al. 2016), Austin's ULL focused on the promises of smart technologies for realising low-carbon transitions and sustainable water management. The ULL project was a collaboration between the City of Austin, a large development company, a non-profit research organisation, a large non-profit environmental organisation, and the University of Texas (UT) with goals to test and demonstrate smart grid infrastructure coupled with smart home technologies and renewable energy generation in "real-world" settings. While the promises of this mode of urban experimentation are considerable, the politics shaping Austin's ULL resulted in a project fuelled by interests in economic growth over radical sustainability.

The data used to construct this case study was collected between March 2014 and May 2016. Primary sources of data included 26 semi-structured, in-depth interviews with government officials, researchers, and representatives of non-profit groups and development companies involved in the project, and observation at various technology showcases, public meetings, and conferences in Austin. Secondary data, including plans and policy documents, news and media, and archival materials, were analyzed through content and discourse analysis. All sources of data were inductively coded to produce themes, then interpreted with theoretical concepts based in governmentality studies and urban governance literatures. While case-studies have intrinsic limitations, the purpose of this paper is to elaborate a conceptual framework that can be expanded, refuted, or supported in further studies of sustainable and smart urban development. This single case study is representative of a broader, diverse and situated phenomenon at the convergence of smart and sustainable cities agendas that I utilise to provide a way of "thinking critically" about urban experimentation.

In the next section, I lay out the theoretical framing for "thinking critically" about urban experimentation which builds from Foucauldian governmentality theory together with geographers work on neoliberal urban governance. I provide a critical view of urban experimentation in order to reclaim a socially just and environmentally sustainable formulation. Then, I turn to the case of Austin, Texas, and show how these logics manifest in a smart and sustainable city project. The case shows how local and state government strategies for economic growth leverage urban experiments to attract technology firms and capital while simultaneously enacting a version of entrepreneurial urbanism that poses as panacea for social and ecological problems. At the same time, the case shows how urban experiments involve socio-technical changes that co-produce norms of conduct for resource consumption enrolled in larger shifts in urban governance, but do not evolve from input from the community itself. In the conclusion, I further elaborate on the possibilities and pitfalls of smart and sustainable urban experimentation for more ecologically and socially just futures.

2. "Thinking critically" about urban experiments

As cities seek to test-out new smart city and sustainability policies, scholars have conceptualised this phenomenon as a mode of governance, a mode of knowledge production and learning, and a form of strategic urban development broadly called *urban experimentation* (Bulkeley, Castán Broto, and Edwards 2014; Caprotti and Cowley 2016; Evans, Karvonen, and Raven 2016; Evans 2016). The logic of experimentation is, quite simply put, to test-out new urban technologies (including smart grids and autonomous vehicles), policies, and partnerships between industry, government, non-profits, and universities (Castán Broto and Bulkeley 2013a; Karvonen and van Heur 2014; McLean, Bulkeley, and Crang 2015). The potential of experimentation is that it will enable learning from interventions in specific urban contexts that enable control and observation of changes over time (Evans 2011; Evans and Karvonen 2014). Knowledge production and diffusion regarding innovative urban policies and accompanying economic stimulus is a central motivation.

While urban experiments are incremental and fragmented projects (although often framed by holistic master plans that suggest controlled and systematic development), they shape local policy and the urban environment itself (Caprotti and Cowley 2016; Cugurullo 2017; Evans 2016). Experiments, more generally, are "purposive and strategic but explicitly seek to capture new forms of learning or experience … they are interventions to try out new ideas and methods in the context of future uncertainties serving to understand how interventions work in practice, in new contexts where they are thought of as innovative" (Castán Broto and Bulkeley 2013a, 93). They offer the "means through which discourses and visions concerning the future of cities are rendered practical, and governable" (Castán Broto and Bulkeley 2013b, 367). Experiments, thus, are also public engagements that aim to persuade audiences, in this case, as to how effective or worthwhile a smart and sustainable city agenda may be. Following this argument, urban experiments are essential elements in constructing political power behind smart and sustainable city projects. While powerful opportunities to introduce alternative logics and models for sustainable urban development, smart and sustainable city experiments too often focus on economic growth and individual responsibility, benefitting well-off households and technology companies that seek profits without concern for social outcomes.

Adding to this existing literature on urban experimentation, this paper presents a critical framework for analyzing urban experiments through a Foucauldian lens of governmentality. The framework asks three questions: (1) What are the dominant motivations for urban experimentation and who stands to benefit? (2) How does urban experimentation shape approaches to sustainability and justice? (3) How do urban experiments engage communities/citizens, and with what implications?

These three questions offer an opportunity for critique and reflexivity in urban experimentation. As experiments grow as a defining feature of urban (environmental) governance, they translate ideas and visions into reality through projects and policies that shape how urban development is conducted. Under neoliberal regimes of urban governance, however, ideas of sustainability become associated with technological fixes often in the form of calculative devices for managing individual resource consumption (such as real-time energy and water monitoring, carbon audits, etc.) that promise to facilitate ecologically sound urban development (Braun 2014). Furthermore, the power of experimentation comes not only from the experience of a single place. Experiments shape the perception of possible urban futures and future action. Experiments can thus be understood as places wherein knowledge claims are deemed credible and authoritative, having broader impacts on agendas for governing cities.

This framework employs a governmentality lens to answer the three questions posed above, responding to three interconnected elements of governmentality analyses: (1) rationalities, (2) techniques, and (3) subjectivities. Governmental rationalities are systems of thinking about the practice of government (who can govern, what governing itself is, and what or who is governed) as a way of making that activity itself practicable (Gordon 1991; Lemke 2001, 2002). Rationalities are formed around problematizations. Foucault was specifically concerned with the problematization of population, for example, and how population created a political necessity and possibility for governmental thought: how to manage and govern populations within a certain territory (Foucault 2009). In the case of smart and sustainable city experimentation, the city and citizens are problematised as unsustainable, inefficient, and in need of economic development. Here, a critical approach asks: What are the dominant motivations for urban experimentation and who stands to benefit?

Techniques refer to the "how" of governing. For Foucault, this included disciplinary techniques, "techniques of power," or power/knowledge "designed to observe, monitor, shape and control the behavior of individuals" in a variety of institutions (Foucault 1977; Gordon 1991, 3). In his later lectures on neoliberalism, Foucault argued that power also operated through forms of "freedom" in neoliberal governmentality that shaped the conduct of individuals through entrepreneurial and competitive mandates (Barry and Osborne 1996; Foucault 2010; Rose 1999). In other words, freedom was conscribed to "self-management" and introduced a new form of "self-regulation" always in relation to a broader set of norms that one either tried to distance themselves from or collapse onto (i.e. self-improvement or responsibility). In smart and sustainable city experimentation, techniques are broad and range from land use planning (Leffers and Ballamingie 2013) to digital maps that produce urban territories (Luque-Ayala and Neves Maia 2018) to urban dashboards that position particular (often contestable) metrics as essential monitors of urban sustainability (Kitchin, Lauriault, and McArdle 2015; Kitchin, Lauriault, and McArdle 2015; Kitchin, Maalsen, and McArdle 2016). Here, a critical approach asks: How does urban experimentation shape approaches to sustainability and justice?

Subjectivities are central to governmental power. For example, Foucault (2010) worked through the political rationality of liberalism and neoliberalism in his lectures, *The Birth of Biopolitics*, explicating the ways in which neoliberal rationality was entwined with market activities and the proliferation of enterprise into the social body, working on and through a particular subject, *homo economicus*. That is, Foucault's interest in power was related to the ways in which it shaped subjects and "proper" conduct. In smart and sustainable city experimentation, we must question how disciplinary power, norms, and other power/knowledge technologies conscribe subject positions as sustainable or not (Gabrys 2014; Sadowski and Pasquale 2015; Vanolo 2013). Here, a critical approach asks: How do urban experiments engage communities/citizens, and with what implications?

Utilising this framework, I argue that smart and sustainable urban experiments as currently developed tend to deepen neoliberal regimes of government marked by tenets of entrepreneurialism and responsibilization. This argument is based on critical readings of the political rationalities of neoliberalism which argue that neoliberalism extends economisation into all aspects of life (Brown 2015). Governmental rationalities are deeply intertwined with the production of subjects. The neoliberal subject, *homo economicus*, is shaped in relation to the "idea and practice of responsibilization – forcing the subject to become a responsible self-investor and self-provider – which reconfigures the correct comportment of the subject from one naturally driven by satisfying interests to one forced to engage in a particular form of self-sustenance that meshes with the morality of the state and the health of the economy" (Brown 2015, 84). In neoliberal versions of urban experimentation, subjectivity is associated with responsibilization for environmental sustainability and economic productivity.

Strategies of neoliberal subjectification are deeply integrated into regimes of urban entrepreneurialism. Urban policy and development in an era marked by a roll-back of state funding has turned towards private capital to fund everyday city services, development projects, and provide "public" goods (Hackworth 2008; Keil 2009). Harvey (1989) described this transition from managerialism (under welfare-state conditions) to entrepreneurialism (under neoliberalism) as a defining feature of urban governance. Here, cities have to brand and market their place in order to attract companies, often leading to lob-sided public-private partnerships, and new forms of real estate speculation (Hall and Hubbard 1996). Cities also utilise development projects to market themselves as sustainable, resilient, or smart in order to further attract investment (Hollands 2008; Long 2016; While, Jonas, and Gibbs 2004) and particular kinds of urban subjects recognised as creative "talent" (Jensen 2007; McLean 2014). Here the focus of smart and sustainable city experimentation takes on an exclusionary logic, as the city becomes marketed as a place for large technology companies to test their products and services, and as some groups of people are privileged over others. This is precisely why a critical framework is necessary for analyses of urban experimentation: it asks critical questions that open smart and sustainable city experimentation up for critical inquiry and reconceptualization.

The next section works through the case of Austin's urban smart grid experimentation described in the Introduction. I demonstrate how this framework for "thinking critically" about urban experimentation can uncover the neoliberal logics and rationalities of an ostensibly sustainable and socially responsible project. The case illustrates how the focus on economic growth, demonstration of technological efficiency, and individual consumption habits limits the power of experimentation for transformative change.

3. Austin's living laboratory for smart technologies

The transition of urban electricity systems towards more sustainable and resilient forms is a vital ingredient in combating climate change. Urban experiments, and urban living labs (ULLs) in particular, have been positioned to address this challenge, often by utilising smart technologies including the "smart grid." The smart grid carries with it the promises of decarbonisation and increased reliability, economic efficiency and new business models, and the ability to harness innovations such as electric vehicles and energy storage through the integration and application of information and communications technologies (ICTs). The smart grid is a sociotechnical intervention that includes more than technologies, but also knowledge networks, user practices, and new business models for energy utilities, data analytics, and grid operations (Luque 2014, 160). Thus, the smart grid carries opportunities for sociotechnical innovations, new relations of production and consumption, and is a tool for generating greater investment and public acceptance for smart city projects.

Austin's ULL for the smart grid, known locally as the Pecan Street Project (PSP), has its roots in the idea of the "technopolis." The technopolis was developed in the thought of Austin's tech "godfather" George Kozmetsky, who helped develop a University of Texas (UT) startup incubator called the Austin Technology Incubator (ATI) in 1989 (Butler 2011). ATI was seen as a way to "future proof" Austin, setting it up to be a leader in the high-tech economy (Butler 2011). ATI has been promoted as an economic engine for the City, propelling startups and spinoff companies while attracting venture capital and talented researchers and workers (Calnan 2014). Partnering with the ATI, City of Austin, Austin Energy, UT researchers, tech companies, and the Environmental Defense Fund, a non-profit research organisation called the Pecan Street Project – later Pecan Street, Inc. – was developed in 2009 to foster, test, and pilot smart grid innovations related to advanced technology, business models, and customer behaviour.

The PSP began as conversation about the "electric internet" between some of Austin government elites, including most prominently several former city council members, UT professors, and a previous mayor pro-tem. Initially described as an "energy internet demonstration project," PSP served to test various socio-technical interventions including electric vehicles and distributed solar, smart home technologies, and advanced smart grid infrastructure in real-world settings. The project aimed to engage environmentally concerned and technologically savvy residents of the newly built Mueller neighbourhood (where the PSP is located) with incentives to adopt electric vehicles, solar panels, home energy management systems (HEMS) and numerous other smart technologies, which then allowed Pecan Street to get considerable participation, conduct research and field trials, and gain access to fine-grained data on energy usage of consumers.

The PSP is located in the Mueller neighbourhood, itself a private-public redevelopment project that commenced in 2004 on a nearly 700-acre defunct airport base just three miles northeast of downtown Austin and the University of Texas. The "greenfield" type development of Mueller enabled construction of smart grid infrastructure from the ground up, which included the development of Austin Energy's smart grid platform (Carvallo and Cooper 2015), new green-built homes, and Pecan Street's own information and communications technologies and smart grid network. Coupled with the various industry or federally supplied smart home technologies, the neighbourhood became a "testbed" for smart technologies and related platforms (Levenda 2018).

3.1. Rationalities and techniques of smart grid experimentation

ULLs test out visions of urban futures that align various actors through reconfigurations of urban infrastructures (Bulkeley, Castan Broto, and Maassen 2014), and at the same time, they provide opportunities to address environmental problems with "testable" solutions in particular places (Bulkeley and Castán Broto 2012; Castán Broto and Bulkeley 2013a). But what are the dominant motivations for Austin's urban experimentation and who stands to benefit? How does this kind of experimentation shape approaches to sustainability and justice? Austin's Pecan Street Project and Mueller redevelopment highlight two salient points about the role of living labs in neoliberal urban politics: the confluence of entrepreneurial and environmental ideals in shaping urban governance (While, Jonas, and Gibbs 2010), and secondly, the mobilisation of these ideas via channels of "fast policy". In the Mueller neighbourhood, demonstration and leadership is centred on the ability to demonstrate a reproducible model, best practices, policy recommendations, or other forms of knowledge that attract attention from other cities and companies. The Mueller neighbourhood is, as a representative of the development company explained:

> A model of urban development, that's part of our vision. We actually learned from Stapleton, and other folks are now learning from us. The other thing that this has done for the City of Austin is that it has become a living lab. (Development Company Representative, Interview, November 2015)

Mueller had to push the threshold of what was prescribed for Austin development projects, and by doing so offered defined spaces wherein new city ordinances could be tested for evaluation and possible future use and adoption in the greater Austin area. The public-private partnership documentation – the Master Development Agreement between the City and developer – gave developers freedom to change the urban form to "new urbanist" styles of development. Yet, this sort of development is indicative of the type of "private over public" partnerships that have become commonplace under neoliberal urbanism, benefiting developers and wealthy Austinites over the people most in need of housing (Weber 2002, 2010). Similarly, these sorts of experimental developments are part and parcel of the acceleration of neoliberal urban policy learning and mobility as cities become more extrospective in search of tools, best practices, reference cases, and knowledge transfer opportunities that will entice private capital (McCann and Ward 2011; Peck and Theodore 2015).

The PSP focused on demonstration of smart grid and smart home technologies as a way to meet federal and local smart grid programme goals, but also as a way to place Austin as the forefront of high-tech, smart city innovations. As a partner, the Environmental Defense Fund (2014) ardently promoted the project:

> The Mueller neighborhood, the locus of Pecan Street, is a laboratory of ideas and technologies that will move the nation's $1.3 trillion electricity market toward a future in which energy is cheap, abundant and clean. If Pecan Street is successful, every neighborhood in America will look like it in 20 years.

EDF mobilised the notion of the laboratory to position the city as a testbed for learning about socio-technical innovations that meet the discursive mandate of triple-bottom-line urban sustainability (Davidson 2010; Gunder and Hillier 2009). Environmental sustainability is leveraged as a marketable idea, one that enhances Austin's competitive advantage. Entrepreneurial urban strategies and policies have utilised sustainability as a discursive move, often with some positive impacts for carbon reduction, and a central focus in order to secure new spaces for development and to neutralise political opposition (Davidson and Gleeson 2014; While, Jonas, and Gibbs 2004). The so-called "new environmental politics of urban development," makes carbon reduction strategies, like the implementation of smart grid interventions and new infrastructures, key to interurban competition and the practices of local governance and planning (Jonas, Gibbs, and While 2011; While, Jonas, and Gibbs 2010).

Furthermore, Austin's Chamber of Commerce, a board member of Pecan Street, has developed strategies to attract startups with potential to receive venture capital for growing their companies. They aim to attract startups working in the spaces of clean energy, creative industries, digital media, and data management, in line with Austin's (highly uneven) creative culture and high-tech economy (Long 2010). The Chamber boasts Austin Energy's commitment to renewable power, the Pecan Street's research potential, and ERCOT's willingness to integrate clean energy companies into their electric grid as key factors for attracting energy companies.[1] Similarly, the University of Texas Clean Energy Incubator, the Clean TX cluster development organisation, and the already large clean tech industry located in Austin are supportive of the growth of the cluster. These technical and economic considerations are only part of the attraction. As one Chamber of Commerce representative explained:

> Clean energy was supported as a recruitment target starting over 10 years ago, supported by Austin Energy. [...] But, we have a very educated population. [...] Obviously quality of life too. We're a blue dot in a red state. (Chamber of Commerce Representative, Interview, November 2015)

The convergence of cultural and economic innovation are positioned as central to the growth of Austin's economy. Austin's brand of urban entrepreneurialism leverages clean energy and clean tech, positioning the local economy as supportive of progressive environmental politics.

This context explains the motivations behind Austin's PSP ULL, and highlights the way ULLs are positioned to promote ecological modernisation. It also shows the importance of ULLs becoming an exemplary "model" that entrepreneurial city leaders can advertise. As former City Council member and Pecan Street President explained, "We intend to make the Mueller neighborhood an example of what modern neighborhoods can accomplish, with smarter energy management, clean energy generation, and advanced system integration [...] the most self-sufficient and energy efficient neighborhood development in the country" (King 2009). The focus on demonstration and leadership has garnered attention from other cities around the world looking to understand how to effectively integrate smart technologies. Austin's Mayor Steve Adler remarked in his 2016 State of the City Address that, "Cities from all over our country and the rest of the world send entire delegations here to troop through our offices in hopes of finding the magic formula written on a white board somewhere. These leaders from other cities ask me what makes Austin so special. I tell them about Barton Springs and how our commitment to our environment became perhaps our most important asset." He continued to explain that what makes Austin special is a commitment to innovation: "We have 7% of the state's population but 30% of the new patents. Austin ranks 8th in the country in venture capital investments. A year ago Forbes put us on a list of five cities poised to be the next Silicon Valley Tech Hub. Whether it's Google's driver-less cars, the Pecan Street Project implementing energy-use technology out at Mueller, or our community's seemingly limitless innovations in the field of breakfast tacos, Austin has become a city where good ideas become real"

(Goudreau 2016). The Mayor's speech centred cultural and economic "innovation" as a primary force behind Austin's growth, and noted the Pecan Street ULL as one exceptional case in order to demonstrate the city's leading position.

Pecan Street has been a vital resource for Austin in this regard, serving as the basis of an "innovation cluster" on smart energy systems (Chamber of Commerce Representative Interview, November 2015). Pecan Street's Dataport offers the world's largest database on customer energy use to university researchers (for free) and technology companies (for a fee). As one Pecan Street representative explained:

> I talk to cities; I talk to for-profit companies. We meet with them and they say, what have you learned, and I'll be happy to tell the for-profit company that is trying to build a product that this is what we've learned, this is what's failed, and this is what's succeeded. [...] We're happy to show off our work [...] My job is to make sure we can get as much data as possible to give to people so they can utilize it and learn from it. (Pecan Street representative, Interview, October 2015)

In this way, living labs, such as Pecan Street's smart grid project can be viewed not only as a place for research and learning, but also as a "theatre of proof" (Simakova 2010; Smith 2009) for ways of configuring smart technologies in urban space to achieve sustainable, low-carbon outcomes. Akin to contemporary practices of the product "launch" in high-tech industries, the "theatre of proof" is described as a situation where an organisation "offers a 'novel' product to 'the market'" (Simakova 2010, 549), and this case that means data about households energy and water consumption, preferences for certain technologies over others, and the efficiency of grid integrated products like electric vehicles and energy storage systems. Here the rationality of experimentation in ULLs is to learn about tech-company-driven agendas for smart technologies, and to gain broader recognition for the city as an innovative space for testing various smart city technologies.

In Austin, for example, having a large public-private redevelopment project provided the opportunity for the PSP to flourish in a community of so-called "early adopters": largely upper-middle class residents that are motivated to save energy or participate in research (Pecan Street representative, Interview, October 2015). Smart grid experimentation in ULLs certainly might yield worthwhile research on the technical limits of the smart grid (such as studies that find how many electric vehicles can be charging in one location without stressing distribution infrastructure), but this approach contributes to the lock-in of particular pathways for smart infrastructure development without broader consideration of the concerns of citizens or the structural limitations to managing energy consumption and production.

Furthermore, while these forms of experimentation may provide some insight into policy pathways for smart and sustainable infrastructures, it also is dominated by the need to attract private capital. Fitting with the dominant form of neoliberal urban governance, governing through experiment aligns with the entrepreneurial role of local governments (Davidson and Gleeson 2014; Hall and Hubbard 1996; MacLeod 2002). As Harvey (1989, 5) argued, with the turn from managerialism to entrepreneurialism in urban governance, "investment increasingly takes the form of a negotiation between international finance capital and local powers doing the best they can to maximize the attractiveness of the local site as a lure for capitalist development." Opening up the city as a test-bed and a demonstration site for new smart technologies provides an opportunity to attract highly mobile capital. However, this may have "splintering" impacts in the city (Graham and Marvin 2001) that create spaces of high-value while simultaneously excluding and marginalising other spaces and communities. Therefore, attention to the rationalities of experimentation are central to understanding their impacts and to understanding who controls and who benefits from these projects.

3.2. Smart consumers

Within the scholarship on ULLs and sustainability transitions, the conception of governing by experiment has been used in a variety of cases to understand the urban governance of sustainability

(Berkhout et al. 2010; Bickerstaff, Hinton, and Bulkeley 2016; Blok and Tschötschel 2015; Bulkeley and Castán Broto 2012; Bulkeley, Castán Broto, and Edwards 2014; Caprotti and Cowley 2016). In ULLs, citizens often become experimental subjects or research participants, such as in the PSP. People are often made to be the object of engagement – the engaged customer, active participant, technology adopter. But how do experiments such as the PSP engage communities/citizens, and with what implications?

More than technological interventions aimed at increasing renewable integration or grid reliability, smart grid experiments attempt to orchestrate and govern energy demand according to particular political-economic rationalities. Smart grid demonstration projects have increasingly used the vocabulary of customer engagement and empowerment (Gangale, Mengolini, and Onyeji 2013). The customer moves beyond the role as a passive consumer and becomes an active participant in the electricity grid with new responsibilities, choices, and opportunities (Naus, van Vliet, and Hendriksen 2015). The growth in attention to demand response, time-of-use pricing, and other "customer side" interventions have been celebrated by utilities and electricity providers as potential opportunities to shave or shift peak demand while increasing customer awareness and engagement. Yet, these opportunities rely on significant changes in energy consumption activities that have not yet been realised (Hargreaves, Nye, and Burgess 2013). Underlying much of these programmes is a conceptualisation of the end-user as a rational economic actor, or what Strengers (2013, 51) calls "resource man" – "a data-driven, information-hungry, technology-savvy home energy manager." These depicted smart end-users are neoliberalized subjects, conscribed by social norms, expected to perform the scripted uses for smart technologies, with the encouragement to act rationally both economically and environmentally.

Tied to the entrepreneurial forms of governance pushed by the policy elite, parallel ideas of entrepreneurialism circulate in discourse to constrain and shape the contours of citizen-consumer subjectivity. In neoliberal urban governance regimes, the emphasis on entrepreneurship shapes expected citizen subject positions. This resonates with Foucault's notion that the role of homo economicus – the consumer subject – is not so much a consumer as a person of "enterprise and production" (Foucault 2008). McNay (2009, 56) explains how neoliberal governmentality is expressed through the concept of "self as enterprise" where individuals are encouraged to view their lives and identities as an enterprise, "understood as a relation to the self based ultimately on a notion of incontestable economic interest." Foucault's investigation into American neoliberalism explains the proliferation of enterprise, entrepreneurship, and policies empowering individuals into the social fabric, and reveals its central importance as a reinforcement of neoliberal subjectivity. Citizens manage themselves and their households according to the dictates of the economy, maximising efficiency over satisfying desires or interests. Instead, efficiency and economic production replace other forms of self-interest to become dominant, aligned with social norms.

The promise of smart grid projects like the PSP relies, in one part, on behavioural changes of users. Here, the expectation is that "smart users" become active participants in the smart grid, performing their part as solar pioneers, eco-energy misers, or flexible energy users adjusting consumption to the dynamics of a time-of-use rate structure. In this sense, smart grid experiments "success" presupposes (rational and individual) market actors who manage their everyday practices in a careful, calculative, and reflexive way. But as the experience in some of the households in Austin's smart grid experiment show, people do not necessarily act "rationally." A Pecan Street representative explained this point directly with a story of a multi-family tenant and research participant whose data profile showed his oven was always on:

> The [resident] goes, 'yeah, my oven is on,' and our technician was like, 'no no no, we *show* that your oven is on' and the [resident] says, 'yeah, my oven's on.' And it turns out this guy just left his oven on all the time. […] We said, it costs a lot of money to you. And the guy had no clue. (Pecan Street Representative, Interview, October 2015).

Similarly, a representative from the Environmental Defense Fund explained that from his research in relation to the smart grid experiment in Austin, people just don't think or care about energy enough to change their behaviour or their practices:

> There is a statistic that is widely quoted that people think about their energy bills and electricity six minutes a year. For most people it's not something that you choose to focus on. One barrier for scaling up demand response and smart technology and that sort of thing is just generating interest. [...] Even if they don't think about it that much, they think about ways to save money, if something is a no-brainer, then you make that choice. (EDF Representative, Interview, October 2015)

As energy researchers engage with smart grid users, they often seem to get dismayed by the irrationality of human behaviour. As the quote above illustrates, the researchers involved with smart grid experiments understand that end-users don't necessarily think about energy very often, but they still feel they can be persuaded economically. This logic is changing the nature of smart grid implementation. Lessons learned from early studies on energy efficiency impacts of smart metres and in-home displays suggest little evidence of sustained behaviour change (Hargreaves, Nye, and Burgess 2010; Hargreaves, Nye, and Burgess 2013). Studies on voluntary demand response and time of use pricing have indicated that these options may work (Dyson et al. 2014; Muratori, Schuelke-Leech, and Rizzoni 2014), but automating decision-making to maximise energy and economic efficiency now is the dominant trend. As one EDF representative explained:

> Using machines and technology doesn't have the human error element, or the human-interest level, so you have these items programmed to be more efficient, and at scale, that will take a lot of the human element of being more efficient with energy out of the equation. It just makes it easier for humans to act with the environment in mind. (EDF Representative, Interview, October 2015)

While automation may provide energy and cost savings for end-users, it also rationalises and normalises the deep integration of smart technologies in everyday life without deliberation over end-users concerns or values (Strengers 2013). This concretises a technological fix for the seeming inflexibility of energy demand and the irrationality of human behaviour, essentially framing energy problems as purely technical ones. But, these problems are more than technical. Energy demand is structured by the rhythms and patterns of everyday life (Walker 2014). Consumption is not for the sake of consumption, but rather for aiding in everyday activities shaped by social norms, habits, economic demands, and other conventions (Shove, Pantzar, and Watson 2012; Shove and Walker 2014). The limitations of the techno-economic approach exemplified by the forms of experimentation discussed in this paper is that the design and implementation of ULLs –whether for smart grid experiments or other purposes – needs to be questioned along axes of social and political concern. And centrally, it should consider how these technological systems lock-in expectations, or technological scripts (Akrich 1995; Gjøen and Hård 2002), that assume particular subject positions and desires without addressing what communities consider important or needed.

4. Experimentation, governance, and smart urban futures

In this paper, I have shown how urban entrepreneurialism and neoliberal governmentality are shaping the design and influencing the proliferation of governance experiments for smart infrastructures in Austin, Texas. I have highlighted how a specific ULL – the Pecan Street smart grid experiment at the Mueller neighbourhood – was constructed materially and discursively as a place of demonstration for public approval and a test-bed for smart technologies. The entrepreneurial push to locate such a project in Austin was spurred by the City's legacy as a technopolis – an economic development strategy that aspires to attract high-tech companies and creative class workers through leveraging the University of Texas and other public resources. Although there are certainly positive pathways for urban experimentation, this case has shown that the role of citizens in determining or influencing the pathways to a smart urban future are limited to very narrow realms of participation through consumption, and it has further shown how these projects seem to only apply to a particular segment of the population – "early adopters" and those with enough wealth to afford to live in places like the Mueller neighbourhood. As such, this paper demonstrates the need for thinking critically about smart and sustainable city experiments along three dimensions: the motivations of experimentation, how experimentation shapes action for sustainability and justice, and the ways communities and citizens are engaged.

With the ever greater entrenchment of smart technologies in the urban environment, greater amounts of data being collected and analyzed, and new programmes for managing and governing urban infrastructures, the smart city creates opportunities for the deepening of surveillance in everyday life (Klauser and Albrechtslund 2014) and provides new avenues for "corporate storytelling" (Söderström, Paasche, and Klauser 2014) to influence entrepreneurial urban governance. This paper has raised several additional critical issues for future scholarship on urban governance at the convergence of smart and sustainable city projects. First, although ULLs or other forms of experimentation promise ways to test-out solutions for urban sustainability, they are largely shaped by existing governance regimes with political economic interests and goals. This may contribute to creating exclusive sustainable enclaves in the city where only small portions of the population benefit. Second, as demonstrations, ULLs are significant opportunities to enrol public support for addressing urban sustainability. Yet, these approaches seem to have a limited approach due to the ways in which citizens can participate in these projects. The Austin case study demonstrates that techno-economic approaches seek to regulate the conduct of individuals through economic incentives, but this approach limits more democratic citizen-led alternatives. This points to the need to reinvigorate governance experimentation with a radically democratic agenda, and that we should take seriously the role that seemingly one-off experiments have for possible co-production of more sustainable and just urban futures.

Note

1. Please see: https://www.austinchamber.com/economic-development/key-industries/clean-energy-power.

Disclosure statement

No potential conflict of interest was reported by the author.

ORCID

Anthony M. Levenda http://orcid.org/0000-0002-3325-8804

References

Akrich, M. 1995. "User Representations: Practices, Methods and Sociology." In *Managing Technology in Society. The Approach of Constructive Technology Assessment*, edited by J. Rip, A. Misa, and T. J. e Schot, 167–184. Pinter. https://halshs.archives-ouvertes.fr/halshs-00081749.

Balta-Ozkan, N., R. Davidson, M. Bicket, and L. Whitmarsh. 2013. "Social Barriers to the Adoption of Smart Homes." *Energy Policy* 63: 363–374. https://doi.org/10.1016/j.enpol.2013.08.043.

Barry, A., and T. Osborne. 1996. *Foucault and Political Reason: Liberalism, Neo-Liberalism, and Rationalities of Government*. Chicago, IL: University of Chicago Press.

Berkhout, F., G. Verbong, A. J. Wieczorek, R. Raven, L. Lebel, and X. Bai. 2010. "Sustainability Experiments in Asia: Innovations Shaping Alternative Development Pathways?" *Environmental Science & Policy* 13 (4): 261–271. https://doi.org/10.1016/j.envsci.2010.03.010.

Bickerstaff, K., E. Hinton, and H. Bulkeley. 2016. "Decarbonisation at Home: The Contingent Politics of Experimental Domestic Energy Technologies." *Environment and Planning A*, 0308518X16653403. https://doi.org/10.1177/0308518X16653403.

Blok, A., and R. Tschötschel. 2015. "World Port Cities as Cosmopolitan Risk Community: Mapping Urban Climate Policy Experiments in Europe and East Asia." *Environment and Planning C: Government and Policy*, 0263774X15614673. https://doi.org/10.1177/0263774X15614673.

Brand, P. 2007. "Green Subjection: The Politics of Neoliberal Urban Environmental Management." *International Journal of Urban and Regional Research* 31 (3): 616–632. https://doi.org/10.1111/j.1468-2427.2007.00748.x.

Braun, B. P. 2014. "A new Urban Dispositif? Governing Life in an age of Climate Change." *Environment and Planning D: Society and Space* 32 (1): 49–64. https://doi.org/10.1068/d4313.

Brown, W. 2015. *Undoing the Demos: Neoliberalism's Stealth Revolution.* Cambridge, MA: MIT Press.

Buchholz, B. M., and Z. Styczynski. 2014. *Smart Grids – Fundamentals and Technologies in Electricity Networks.* Heidelberg: Springer.

Buck, N. T., and A. While. 2015. "Competitive Urbanism and the Limits to Smart City Innovation: The UK Future Cities Initiative." *Urban Studies,* 0042098015597162. https://doi.org/10.1177/0042098015597162.

Bulkeley, Harriet, and V. Castán Broto. 2012. "Urban Experiments and Climate Change: Securing Zero Carbon Development in Bangalore." *Contemporary Social Science* (May 2014): 1–22. https://doi.org/10.1080/21582041.2012.692483.

Bulkeley, Harriet, V. Castán Broto, and G. A. S. Edwards. 2014. *An Urban Politics of Climate Change: Experimentation and the Governing of Socio-Technical Transitions.* London: Routledge.

Bulkeley, H., V. Castan Broto, and A. Maassen. 2014. "Low-carbon Transitions and the Reconfiguration of Urban Infrastructure." *Urban Studies* 51 (7): 1471–1486. https://doi.org/10.1177/0042098013500089.

Butler, J. S. 2011. Kozmetsky was Austin's tech godfather, says UT official. *Austin Statesman,* November 5. http://www.statesman.com/news/opinion/kozmetsky-was-austin-tech-godfather-says-official/RZeoshYq2JhrgnY6HRFMhl/.

Calnan, C. 2014. "Austin Technology Incubator Sparks $880M Economic Impact, Report Says." *Austin Business Journal,* March 5. https://www.bizjournals.com/austin/blog/techflash/2014/03/austin-technology-incubator-sparks-880m-economic.html.

Caprotti, F., and R. Cowley. 2016. "Interrogating Urban Experiments." *Urban Geography,* 1–10. https://doi.org/10.1080/02723638.2016.1265870.

Cardullo, P., R. Kitchin, and C. Di Feliciantonio. 2017. "Living Labs and Vacancy in the Neoliberal City." *Cities* 73: 44–50. https://doi.org/10.1016/j.cities.2017.10.008.

Carvallo, A., and J. Cooper. 2015. *The Advanced Smart Grid: Edge Power Driving Sustainability, Second Edition.* Boston: Artech House.

Castán Broto, V., and H. Bulkeley. 2013a. "A Survey of Urban Climate Change Experiments in 100 Cities." *Global Environmental Change : Human and Policy Dimensions* 23 (1): 92–102. https://doi.org/10.1016/j.gloenvcha.2012.07.005.

Castán Broto, V., and H. Bulkeley. 2013b. "Maintaining Climate Change Experiments: Urban Political Ecology and the Everyday Reconfiguration of Urban Infrastructure." *International Journal of Urban and Regional Research* 37 (6): 1934–1948. https://doi.org/10.1111/1468-2427.12050.

Colding, J., and S. Barthel. 2017. "An Urban Ecology Critique on the "Smart City" Model." *Journal of Cleaner Production* 164: 95–101. https://doi.org/10.1016/j.jclepro.2017.06.191.

Cugurullo, F. 2017. "Exposing Smart Cities and Eco-Cities: Frankenstein Urbanism and the Sustainability Challenges of the Experimental City." *Environment and Planning A.* 0308518X1773853. https://doi.org/10.1177/0308518X17738535.

Davidson, M. 2010. "Sustainability as Ideological Praxis: The Acting out of Planning's Master-Signifier." *City* 14 (4): 390–405. https://doi.org/10.1080/13604813.2010.492603.

Davidson, K., and B. Gleeson. 2014. "The Sustainability of an Entrepreneurial City?" *International Planning Studies* (April 2014): 1–19. https://doi.org/10.1080/13563475.2014.880334.

Dyson, M. E. H., S. D. Borgeson, M. D. Tabone, and D. S. Callaway. 2014. "Using Smart Meter Data to Estimate Demand Response Potential, with Application to Solar Energy Integration." *Energy Policy* 73: 607–619. https://doi.org/10.1016/j.enpol.2014.05.053.

Evans, J. P. 2011. "Resilience, Ecology and Adaptation in the Experimental City." *Transactions of the Institute of British Geographers* 36 (2): 223–237. https://doi.org/10.1111/j.1475-5661.2010.00420.x.

Evans, Joshua. 2016. "Trials and Tribulations: Problematizing the City Through/as Urban Experimentation." *Geography Compass* 10 (10): 429–443. https://doi.org/10.1111/gec3.12280.

Evans, James, R. Jones, A. Karvonen, L. Millard, and J. Wendler. 2015. "Living Labs and co-Production: University Campuses as Platforms for Sustainability Science." *Current Opinion in Environmental Sustainability* 16: 1–6. https://doi.org/10.1016/j.cosust.2015.06.005.

Evans, James, and A. Karvonen. 2014. "'Give Me a Laboratory and I Will Lower Your Carbon Footprint!' — Urban Laboratories and the Governance of Low-Carbon Futures." *International Journal of Urban and Regional Research* 38 (2): 413–430. https://doi.org/10.1111/1468-2427.12077.

Evans, James, A. Karvonen, and R. Raven. 2016. *The Experimental City.* London: Routledge.

Foucault, M. 1977. *Discipline and Punish: The Birth of the Prison.* New York: Vintage Books.

Foucault, M. 2009. *Security, Territory, Population: Lectures at the Collège de France 1977–1978.* New York: Palgrave Macmillan.

Foucault, M. 2010. *The Birth of Biopolitics: Lectures at the Collège de France, 1978–1979.* New York: Palgrave Macmillan.

Gabrys, J. 2014. "Programming Environments: Environmentality and Citizen Sensing in the Smart City." *Environment and Planning D: Society and Space* 32 (1): 30–48. https://doi.org/10.1068/d16812.

Gangale, F., A. Mengolini, and I. Onyeji. 2013. "Consumer Engagement: An Insight from Smart Grid Projects in Europe." *Energy Policy* 60: 621–628. https://doi.org/10.1016/j.enpol.2013.05.031.

Gjøen, H., and M. Hård. 2002. "Cultural Politics in Action: Developing User Scripts in Relation to the Electric Vehicle." *Science, Technology, & Human Values* 27 (2): 262–281.

Gordon, C. 1991. "Governmental Rationality: An Introduction." In *The Foucault Effect: Studies in Governmentality: With Two Lectures by and an Interview with Michel Foucault*, edited by M. Foucault, G. Burchell, and P. Miller, 1–51. Chicago: University of Chicago Press.

Goudreau, A. 2016. Austin Mayor's State of the City: Great Cities do Big Things, February 17. Accessed August 6, 2017. http://www.kvue.com/news/local/austin-mayors-state-of-the-city-great-cities-do-big-things/45204038.

Graham, S., and S. Marvin. 2001. *Splintering Urbanism: Networked Infrastructures, Technological Mobilities and the Urban Condition*. London and New York: Routledge.

Gunder, M., and J. Hillier. 2009. *Planning in Ten Words Or Less: A Lacanian Entanglement with Spatial Planning*. London: Routledge.

Hackworth, J. 2008. "The Durability of Roll-Out Neoliberalism Under Centre-Left Governance: The Case of Ontario's Social Housing Sector." *Studies in Political Economy* 81 (1): 7–26. https://doi.org/10.1080/19187033.2008.11675071.

Hall, T., and P. Hubbard. 1996. "The Entrepreneurial City: New Urban Politics, New Urban Geographies?" *Progress in Human Geography* 20 (2): 153–174. https://doi.org/10.1177/030913259602000201.

Hargreaves, T., M. Nye, and J. Burgess. 2010. "Making Energy Visible: A Qualitative Field Study of how Householders Interact with Feedback from Smart Energy Monitors." *Energy Policy* 38 (10): 6111–6119. https://doi.org/10.1016/j.enpol.2010.05.068.

Hargreaves, T., M. Nye, and J. Burgess. 2013. "Keeping Energy Visible? Exploring how Householders Interact with Feedback from Smart Energy Monitors in the Longer Term." *Energy Policy* 52: 126–134. https://doi.org/10.1016/j.enpol.2012.03.027.

Harvey, D. 1989. "From Managerialism to Entrepreneurialism: The Transformation in Urban Governance in Late Capitalism." *Geografiska Annaler: Series B, Human Geography* 71 (1): 3–17.

Hollands, R. G. 2008. "Will the Real Smart City Please Stand up?" *City* 12 (3): 303–320. https://doi.org/10.1080/13604810802479126.

Höjer, M., and J. Wangel. 2015. "Smart Sustainable Cities: Definition and Challenges." In *ICT Innovations for Sustainability*, edited by L. M. Hilty, and B. Aebischer, 333–349. Springer International Publishing. http://link.springer.com/chapter/10.1007/978-3-319-09228-7_20.

IBM. 2015. IBM - Smarter Cities - building and carrying out ways for a city to realize its full potential, July 23. Accessed August 31, 2015. http://www.ibm.com/smarterplanet/us/en/smarter_cities/overview/index.html.

Jensen, O. B. 2007. "Culture Stories:Understanding Cultural Urban Branding." *Planning Theory* 6 (3): 211–236. https://doi.org/10.1177/1473095207082032.

Jonas, a. E. G., D. Gibbs, and a. While. 2011. "The New Urban Politics as a Politics of Carbon Control." *Urban Studies* 48 (12): 2537–2554. https://doi.org/10.1177/0042098011411951.

Karvonen, A., and B. van Heur. 2014. "Urban Laboratories: Experiments in Reworking Cities." *International Journal of Urban and Regional Research* 38 (2): 379–392. https://doi.org/10.1111/1468-2427.12075.

Keil, R. 2009. "The Urban Politics of Roll-with-it Neoliberalization." *City* 13 (2–3): 230–245. https://doi.org/10.1080/13604810902986848.

King, M. 2009. Pecan Street Project. *Austin Chronicle*, December 4. https://www.austinchronicle.com/news/2009-12-04/924767/.

Kitchin, R., T. P. Lauriault, and G. McArdle. 2015. "Knowing and Governing Cities Through Urban Indicators, City Benchmarking and Real-Time Dashboards." *Regional Studies, Regional Science* 2 (1): 6–28. https://doi.org/10.1080/21681376.2014.983149.

Kitchin, R., S. Maalsen, and G. McArdle. 2016. "The Praxis and Politics of Building Urban Dashboards." *Geoforum; Journal of Physical, Human, and Regional Geosciences* 77: 93–101. https://doi.org/10.1016/j.geoforum.2016.10.006.

Klauser, F. R., and A. Albrechtslund. 2014. "From Self-Tracking to Smart Urban Infrastructures: Towards an Interdisciplinary Research Agenda on Big Data." *Surveillance & Society* 12 (2): 273–286.

Leffers, D., and P. Ballamingie. 2013. "Governmentality, Environmental Subjectivity, and Urban Intensification." *Local Environment* 18 (2): 134–151. https://doi.org/10.1080/13549839.2012.719016.

Lemke, T. 2001. "'The birth of bio-politics': Michel Foucault's lecture at the Collège de France on neo-liberal governmentality." *Economy and Society* 30 (2): 190–207. https://doi.org/10.1080/03085140120042271.

Lemke, T. 2002. "Foucault, Governmentality, and Critique." *Rethinking Marxism* 14 (3): 49–64. https://doi.org/10.1080/089356902101242288.

Long, J. 2010. *Weird City: Sense of Place and Creative Resistance in Austin, Texas*. Austin, TX: University of Texas Press.

Long, J. 2016. "Constructing the Narrative of the Sustainability fix: Sustainability, Social Justice and Representation in Austin TX." *Urban Studies* 53 (1): 149–172. https://doi.org/10.1177/0042098014560501.

Levenda, A. M. 2018. "Mobilizing Smart Grid Experiments: Policy Mobilities and Urban Energy Governance." *Environment and Planning C: Politics and Space*. Online First. http://dx.doi.org/10.1177/2399654418797127

Luque, A. 2014. "The Smart Grid and the Interface Between Energy, ICT, and the City." In *Urban Retrofitting for Sustainability: Mapping the Transition to 2050*, edited by T. Dixon, M. Eames, M. Hunt, and S. Lannon. New York: Routledge.

Luque-Ayala, A., and F. Neves Maia. 2018. "Digital Territories: Google Maps as a Political Technique in the re-Making of Urban Informality." *Environment and Planning D: Society and Space*. 1–19. https://doi.org/10.1177/0263775818766069.

MacLeod, G. 2002. "From Urban Entrepreneurialism to a "Revanchist City"? On the Spatial Injustices of Glasgow's Renaissance." *Antipode* 34 (3): 602–624. https://doi.org/10.1111/1467-8330.00256.

McCann, E., and K. Ward. 2011. *Mobile Urbanism: Cities and Policymaking in the Global Age*. Minneapolis: University of Minnesota Press.

McLean, H. 2014. "Digging Into the Creative City: A Feminist Critique: Digging Into the Creative City." *Antipode* 46 (3): 669–690. https://doi.org/10.1111/anti.12078.

McLean, A., H. Bulkeley, and M. Crang. 2015. "Negotiating the Urban Smart Grid: Socio-Technical Experimentation in the City of Austin." *Urban Studies*, 0042098015612984. https://doi.org/10.1177/0042098015612984.

McNay, L. 2009. "Self as Enterprise Dilemmas of Control and Resistance in Foucault's The Birth of Biopolitics." *Theory, Culture & Society* 26 (6): 55–77. https://doi.org/10.1177/0263276409347697.

Moloney, S., and R. Horne. 2015. "Low Carbon Urban Transitioning: From Local Experimentation to Urban Transformation?" *Sustainability* 7 (3): 2437–2453. https://doi.org/10.3390/su7032437.

Moloney, S., and Y. Strengers. 2014. "'Going Green'?: The Limitations of Behaviour Change Programmes as a Policy Response to Escalating Resource Consumption." *Environmental Policy and Governance* 24 (2): 94–107. https://doi.org/10.1002/eet.1642.

Muratori, M., B.-A. Schuelke-Leech, and G. Rizzoni. 2014. "Role of Residential Demand Response in Modern Electricity Markets." *Renewable and Sustainable Energy Reviews* 33: 546–553. https://doi.org/10.1016/j.rser.2014.02.027.

Naus, J., B. J. M. van Vliet, and A. Hendriksen. 2015. "Households as Change Agents in a Dutch Smart Energy Transition: On Power, Privacy and Participation." *Energy Research & Social Science* 9: 125–136. https://doi.org/10.1016/j.erss.2015.08.025.

Peattie, K. 2010. "Green Consumption: Behavior and Norms." *Annual Review of Environment and Resources* 35 (1): 195–228. https://doi.org/10.1146/annurev-environ-032609-094328.

Peck, J., and N. Theodore. 2015. *Fast Policy: Experimental Statecraft at the Thresholds of Neoliberalism*. Minneapolis: University of Minnesota Press.

Rose, N. S. 1999. *Powers of Freedom: Reframing Political Thought. Cambridge, United Kingdom*. New York, NY: Cambridge University Press.

Rosol, M., V. Béal, and S. Mössner. 2017. "Greenest Cities? The (Post-)Politics of new Urban Environmental Regimes." *Environment and Planning A: Economy and Space* 49 (8): 1710–1718. https://doi.org/10.1177/0308518X17714843.

Sadowski, J., and F. Pasquale. 2015. "The Spectrum of Control: A Social Theory of the Smart City." *First Monday* 20 (7), http://firstmonday.org/ojs/index.php/fm/article/view/5903.

Shove, E., M. Pantzar, and M. Watson. 2012. *The Dynamics of Social Practice: Everyday Life and how it Changes*. London: SAGE Publications.

Shove, E., and G. Walker. 2014. "What is Energy for? Social Practice and Energy Demand." *Theory, Culture & Society* 31 (5): 41–58.

Simakova, E. 2010. "RFID "Theatre of the Proof": Product Launch and Technology Demonstration as Corporate Practices." *Social Studies of Science* 40 (4): 549–576. https://doi.org/10.1177/0306312710365587.

Smith, W. 2009. "Theatre of Use: A Frame Analysis of Information Technology Demonstrations." *Social Studies of Science* 39 (3): 449–480. https://doi.org/10.1177/0306312708101978.

Soneryd, L., and Y. Uggla. 2015. "Green Governmentality and Responsibilization: new Forms of Governance and Responses to 'Consumer Responsibility.'." *Environmental Politics* 24 (6): 913–931. https://doi.org/10.1080/09644016.2015.1055885.

Söderström, O., T. Paasche, and F. Klauser. 2014. "Smart Cities as Corporate Storytelling." *City* 18 (3): 307–320. https://doi.org/10.1080/13604813.2014.906716.

Strengers, Y. 2013. *Smart Energy Technologies in Everyday Life: Smart Utopia?* London: Palgrave Macmillan.

Vanolo, A. 2013. "Smartmentality: The Smart City as Disciplinary Strategy." *Urban Studies*, 0042098013494427. https://doi.org/10.1177/0042098013494427.

Voytenko, Y., K. McCormick, J. Evans, and G. Schliwa. 2016. "Urban Living Labs for Sustainability and low Carbon Cities in Europe: Towards a Research Agenda." *Journal of Cleaner Production* 123: 45–54. https://doi.org/10.1016/j.jclepro.2015.08.053.

Wakefield, S., and B. Braun. 2014. "Guest EditorialGoverning the Resilient City." *Environment and Planning D: Society and Space* 32 (1): 4–11. https://doi.org/10.1068/d3201int.

Walker, G. 2014. "The Dynamics of Energy Demand: Change, Rhythm and Synchronicity." *Energy Research & Social Science* 1: 49–55. https://doi.org/10.1016/j.erss.2014.03.012.

Weber, R. 2002. "Extracting Value from the City: Neoliberalism and Urban Redevelopment." *Antipode* 34 (3): 519–540. https://doi.org/10.1111/1467-8330.00253.

Weber, R. 2010. "Selling City Futures: The Financialization of Urban Redevelopment Policy." *Economic Geography* 86 (3): 251–274. https://doi.org/10.1111/j.1944-8287.2010.01077.x.

While, A., A. E. G. Jonas, and D. Gibbs. 2004. "The Environment and the Entrepreneurial City : Searching for the Urban ` Sustainability Fix ' in Manchester and Leeds *." *International Journal of Urban and Regional Research* 28 (3): 549–569.

While, A., A. E. G. Jonas, and D. Gibbs. 2010. "From Sustainable Development to Carbon Control: eco-State Restructuring and the Politics of Urban and Regional Development." *Transactions of the Institute of British Geographers* 35 (1): 76–93. https://doi.org/10.1111/j.1475-5661.2009.00362.x.

The promise of smart grids

Heather Lovell

ABSTRACT
It is the promise of smart grids – their anticipated role in meeting economic, social, environmental policy objectives – that is driving action on smart grids worldwide, while the reality is rather more messy. This paper is about the implementation of smart grids in Australia, and examines the degree to which environmental and social promises have materialised (or not) within two large energy smart grid initiatives undertaken in the period 2009–2014: the federal government-sponsored *Smart Grid Smart City* Program and the State of Victoria's *Advanced Metering Infrastructure* Program. The analysis draws on a governmentality approach to examine how the promise of smart grids has not for the most part been delivered, concentrating in particular on how new digital technologies have not "behaved" in the way originally planned. Within a governmentality framework, it is generally assumed that technologies work to support government programmes, to accomplish governance. But growing evidence points to smart grid technologies undermining the promise of smart grids. Such a finding stands at odds with the assumption in governmentality about technologies doing work in consort with rationalities of government.

1. Introduction

With the roll-out of its ambitious National Broadband Network, Australia has pioneered experimentation with how new fast broadband-enabled capabilities in information technology and "big data" could herald a shift in the design and management of traditional utility infrastructures (DEWHA 2009b; KEMA 2013). It has also positioned itself as a world-leader in developing the first set of technical standards for smart grids (Standards Australia AS 5711 *Smart Grids Vocabulary*). As such, Australia provides an ideal environment for empirical research to compare the promise of smart grids with their actual implementation. In this paper, empirical research on two Australian energy smart grid initiatives – the *Smart Grid Smart City* (SGSC) Program and the State of Victoria's *Advanced Metering Infrastructure* (AMI) Program – provides insights into how and why the high-tech data-led intelligence of smart grids has encountered problems at the point of implementation. The SGSC Program is an urban initiative, centred on the cities of Sydney and Newcastle in New South Wales, although also deliberately including some in rural areas. The Victorian AMI Program, in contrast, is not specifically urban – it covered the whole of the state. Analysis focuses on social and environmental equity issues, and especially how attention to equity changed over time, in large part because of unexpected issues with smart grid technologies as they were implemented and used.

In Australia, as elsewhere, smart grid initiatives have to date largely focused on the energy sector. They are an example of policy initiative that has promised a lot, and that there has been much excitement about. For example:

The optimal deployment of smart grids holds significant potential for the management of many of the challenges confronting the electricity supply chain in Australia. (Lazar and McKenzie 2012, 1); and

Smart grids represent the cutting edge of energy efficient technologies, applied in energy production, distribution and householder use, a frontier the Australian Government is committed to exploring quickly and strategically as we move to a low-carbon future. (Australian Government Minister for the Environment, Heritage and the Arts, cited in DEWHA 2009b, 4)

There is much in keeping with other accounts of energy and infrastructure governance (Graham and Marvin 2001, Hughes 1983; Mitchell 2008; Nye 1992), in particular with regard to findings about the difficulties of governing fast-changing technological sectors, with scholars considering issues such as the potential for unforeseen disruption, the politics of technology, and maintaining democracy (Winner 1977; Zimmerman 1995). Energy is an area that has also been studied to better understand the relationship between society and technological innovation, using Science and Technology Studies conceptual frameworks such as actor-network theory (see, for example, Akrich 1992) and sociotechnical transitions and "lock-in" (see, for example, Turnheim and Geels 2012; Unruh 2002). Whilst these are potentially valuable conceptual frameworks to apply here, in order to analyse the two Australian smart grid cases, instead in this paper I concentrate on the concept of governmentality, because of governmentality's particular attention to the relationship between policy rhetoric or rationalities ("promises") and the substance of policies – the "technologies" of government. Specifically, I wish to explore governmentality's assumption that technologies necessarily work to "accomplish governance" (Dean 1999, 212).

The paper is structured as follows. First, a summary of the methodology is given. Second, a brief overview of recent social science analysis of smart grids is provided, summarising the main ways in which smart grids have been conceptualised. This literature review provides a more general overview of wider conceptual themes and issues that scholars working on smart grids have attended to, before focusing on governmentality. Third, what has happened over time with the two Australian smart grid case studies is examined, from the initial optimism and ambitious objectives about what they would achieve, to the reality of the more circumscribed actual benefits during and after implementation. The empirical focus is on social and environmental equity objectives and their evolution over time: with the Victorian AMI Program at the outset equity issues were not a large component of the programme, but they did become more important as implementation progressed; in contrast with SGSC environmental equity was a central focus of the programme at the outset, but this declined significantly over time. Fourth, the paper concludes with a discussion of policy retreat and the role of technology, with a call for greater consideration of the ways in which smart grid technologies have undermined or supported smart grid policy initiatives, and the insights provided by a governmentality conceptual lens, with a focus on technology, are summarised.

2. Methodology

The paper is based on research conducted in Australia in the period 2015–2017 on two large smart grid initiatives – the Victorian AMI Program and SGSC. The empirical research involved 37 interviews with actors from across the public and private sectors. The selection of interviewees was on a "whole of population" basis (i.e. there was no sampling): all senior people involved in the two projects were approached with a request for an interview. The large majority interviews were in person, with the remainder taking place via video-phone or telephone. Interviews lasted in the range of 20–90 minutes, with most just under an hour in duration. Permission was asked to record the interview, and the digital recordings were transcribed and coded using the software programme N-Vivo. Analysis and coding of the interview material were completed using an inductive, grounded theory approach. Secondary analysis of the two cases involved an extensive review of policy documents and website material, with all material relevant to the two smart grid initiatives reviewed and analysed.

3. Conceptualising smart grids

Smart grids to date have been conceptualised using a number of different theoretical frameworks. Here, the concept of governmentality is used to explore how smart grids were originally conceived and how their implementation has proceeded, but first, a brief broader review of social science scholarship on smart grids and smart cities is provided, in order to situate how – as a relatively new phenomenon – they have been understood.

3.1. Social science smart grid research themes

First, there has been a strong empirical focus on urban areas within smart grids social science research, in particular large or "world cities" (de Jong et al. 2015 Gabrys 2014; Hollands 2008; Luque-Ayala and Marvin 2016; Viitanen and Kingston 2014). This urban focus reflects the policy and practitioner dominance of the smart agenda within cities, whereby "smartness" has become synonymous with urban localities. Urban areas are also viewed as "hotspots" of innovation where the capital required for smart grids (finance, human resources, etc.) is present, and where there is a confluence of dense utility infrastructures (Graham and Marvin 2001; Karvonen and van Heur 2014; Morgan 2004).

Second, there has been close attention to the discourse of "smart" and "the smart city". In general terms, "smart" is defined as about the integration of new digital capabilities into existing utility infrastructures, but there are multiple terms in use: smart cities, smart grids, intelligent networks and so on. De Jong et al. caution that " ... there is ample potential for terminological fuzziness, or even confusion, unless these terms can be clarified as distinct conceptual categories and in relation to specific applications" (2015, 26). De Jong et al. analyse the use of 12 popular terms within the academic literature on sustainable urbanisation: "sustainable city", "eco city", "low carbon city", "liveable city", "green city", "smart city", "digital city", "ubiquitous city", "intelligent city", "information city", "knowledge city" and "resilient city". There are acknowledged tensions between the two agendas – smart and sustainability – embedded within these terms, as examined in this Special Issue.

Other scholars have focused on a more profound lack of clarity about what "smart" means in practice, i.e. how it has materialised and translated from a vision into something implemented, as Hollands explains " ... the disjuncture between image and reality here may be the real difference between a city actually being intelligent, and it simply lauding a smart label" (2008, 305). This "disjuncture" between discourse or image and reality is strong guiding theme for the empirical analysis below, where it is demonstrated how much of the original promise of smart grids has not eventuated in Australia.

Third, there has been close attention to smart grids as an expression of neoliberalism, examining the private sector power, motives and interests in smart grids, and seeing " ... smart cities ... as marketplaces for technology products and services" (Viitanen and Kingston 2014, 804) and how " ... the technological smart city becomes a smokescreen for ushering in the business-dominated informational city" (Hollands 2008, 311). Such analysis adopts a critical stance to the "smart" ideology, critiquing what is viewed as its core focus on economic efficiency, market function and business opportunities. A number of important issues have been raised in this neoliberal critique about the social and equity effect of smart cities, including:

> ... the splintering effects of the informational city, the limits of urban entrepreneurialism, problems created by the creative classes for local communities, including deepening social inequality and urban gentrification, not to mention the conflict between environment sustainability and economic growth. (Hollands 2008, 311)

Identifying and unpacking these social justice issues from the smart grids optimistic (and typically business orientated) policy discourse has been a central plank of social science smart grids research to date.

3.2. Governmentality

Governmentality has been a popular and productive theoretical lens applied to the study of smart grids and smart cities (Bulkeley, Powells, and Bell 2016; Gabrys 2014; Klauser, Paasche, and Söderström 2014; Luque-Ayala and Marvin 2016; McGuirk, Bulkeley, and Dowling 2014), and is employed here as the main theoretical framework guiding analysis, because of its interest in the relationship between government discursive frames or "rationalities" and governmental "technologies". This makes it ideally suited to analysis of the smart grid empirical cases examined below, which feature strong policy promises involving the positioning of smart grids as a solution to a number of policy problems, effected through a range of smart grid technologies. Furthermore, there is an opportunity to closely examine the assumption amongst governmentality scholars that governmental technologies act in ways to support government rationalities.

The term "governmentality" is closely associated with the work of Michel Foucault, who first used the term in his lectures on government at the Collège de France in the late 1970s, and has been taken in various directions by subsequent scholars (Backstrand and Lovbrand 2006; Dean 1999; Li 2007; Power 1994). Governmentality is a huge field of scholarship, and it is not my intention to provide a full and comprehensive review here, but rather to briefly introduce key aspects of governmentality, and then focus on scholars who have specifically examined smart grids using a governmentality framework.

Dean (1999, 209) defines governmentality broadly as "How we think about governing others and ourselves in a wide variety of contexts", with this thinking described as the "rationality" or "mentality" of government. Thus governmentality attends to the relationship between discourse and practice as a way of better understanding how governing is done, and how governance objectives are actually achieved. A core idea of governmentality is that rationalities are informed by, and inform, the "technologies" of government – standards, technical kit, accounting tools, etc. – the day-to-day practices and routine activities that are used to translate policy ideas into practice (Higgins and Larner 2010; Lansing 2011; Li 2007). This is what Li refers to as "rendering technical" – a set of practices:

> concerned with representing "the domain to be governed as an intelligible field with specifiable limits and particular characteristics … defining boundaries, rendering that within them visible, assembling information about that which is included and devising techniques to mobilize the forces and entities thus revealed." (Li 2007, 7; quoting Rose 1999, 52)

and as Dean further explains:

> The analysis of government is concerned with thought as it becomes linked to and is embedded in technical means for the shaping and reshaping of conduct and in practices and institutions. *Thus to analyse mentalities of government is to analyse thought made practical and technical.* (Dean 1999, 18, emphasis added)

Notably, Dean defines the techniques or technologies of government as "The diverse and heterogeneous means, mechanisms and instruments *through which governing is accomplished*" (Dean 1999, 212, emphasis added); with the implication that the technologies of government are acting in ways that support policy objectives.

Thus, for smart grids, a governmentality approach usefully focuses our attention on not just the discourse of smart grids and its resonance and appeal, but also the technical apparatus of smart grids – the digital meters, communications systems, sensors and so on – that populate these discourses. As applied to smart grids and smart cities, a govermentality framework has been used thus far to explore the changing logic or political rationality of governing smart grids (Bulkeley, Powells, and Bell 2016), the effect of smart initiatives on urban citizenship (Gabrys 2014), and the smart grid control room (Luque-Ayala and Marvin 2016), amongst other contributions (see Klauser, Paasche, and Söderström 2014; McGuirk, Bulkeley, and Dowling 2014). For example, Bulkeley, Powells, and Bell (2016) use the concept of governmentality in combination with social practice theory to understand smart grids as a form of political rationality, drawing on empirical research with households with rooftop solar photovoltaics in the north of England, some of which are being encouraged to use new technical configurations in the home so that they export less electricity to the grid (and thereby avoiding voltage fluctuations which the local utility has to manage), instead storing or using it directly within the home. The authors argue that in this instance smart grids are not so much a technical phenomena, but instead represent a more fundamental way of rethinking utility infrastructure provision – a change in the rationality of energy policy, which embraces "bottom up" household changes in energy technologies and practices:

> The smart grid is not a set of policies to be implemented or structures to be built, but rather a governmental pro-gramme … The work of establishing the smart grid therefore requires particular ways of thinking about what the electricity problem is and what desirable electricity futures might involve. This in turn involves establishing the techniques, artefacts and devices through which particular forms of electricity provision that accord with this dynamic of problematisation and improvement can be maintained (and are contested). (Bulkeley, Powells, and Bell 2016, 11)

Although there is an openness here to new energy "techniques, artefacts and devices" being con-tested by households, the central thrust of the analysis – common to other governmentality studies – is about the rationality or governmental programme framing the techniques and devices, with less attention to how technologies "act back", and in turn shape smart grid rationalities. But for Gabrys (2014, 32), in contrast, such an approach is more central, wherein " … the smart city is an ever-elusive project to be realized". Gabrys draws specifically on the notion of "environmentality", defined as " … the distribution of governance within and through environments and environmental technologies" (2014, 30), and uses an empirical case of the Connected Sustainable Cities project, a partnership between MIT and CISCO, applied in eight international cities, to attend closely to the unravellings and tensions that became apparent as the project evolved, noting how:

> Smart-city design proposals on one level establish propositions and programs for how computational urbanisms are to operate; but on another level, *programs never go according to plan and are never singularly enacted.* Envir-onmentality might be advanced by considering smart cities not as the running of code in a command-and-control logic of governing space but as the multiple, iterative, and even faltering materializations of imagined and lived computational urbanisms. (Gabrys 2014, 37, emphasis added)

Thus, as Gabrys and others (see, for example, Hollands 2008) have usefully shown, there are instances where smart grid technologies have not only failed to support but have actually undermined the opti-mistic discourse about smart grids (the "promise"). In the section below, I explore whether this has also been the case in Australia, examining two large smart grid programmes that were implemented at the turn of the century.

4. Australian smart grid programmes

The analysis in this section examines in detail the Australian federal government's SGSC Program (2010–2014) and the State of Victoria's AMI Program (2009–2013). Particular attention is paid to what these programmes promised to deliver at the outset, in the planning stages, compared with what happened subsequently, when implemented. Analysis focuses on social and environmental equity issues, and especially how attention to these issues changed over time, and the role of smart grid technologies therein. In the case of the Victorian AMI Program, at the outset equity issues were not present, but they became a bigger component as implementation occurred and the social equity implications of the metering programme became apparent. In contrast, with SGSC environmental equity was a big feature of the programme at the outset, but this focus declined over time.

4.1. The Victorian AMI Program

The Victorian AMI Program involved the mandatory implementation of digital "advanced" meters to all households and businesses in the Australian State of Victoria. The Victorian government set the policy framework, and the actual work of implementation was done by the electricity distribution companies, of which there are five in Victoria. The AMI Program was given state government approval in 2006 and commenced in 2009, in anticipation at the time of the rest of the National Electricity Market (NEM) in Australia following suit. Discussions around changes to electricity metering in Victoria actually started several years before: there was agreement in 2004 on a programme to replace traditional meters with interval meters on a gradual (non-mandatory) basis (ESC 2004). But, after a number of studies and further consideration (see CRA International and Impaq 2005), it was decided to modify the interval metering programme to a smart or advanced metering programme, i.e. to install meters incorporating two-way communications, and to do this more quickly, with a mandatory "accelerated roll-out" (DPI 2007, 7). In the period 2009–2013, 2.8 million advanced meters were installed on this basis in 93% of homes and small businesses across Victoria in both urban and rural areas (VAGO 2015). This involved removing the old mechanical meter in each property and replacing it with a digital advanced meter. Customers were charged directly for the new meters, with Victorian households paying on average $760 extra on their bills in the period 2010–2015 because of additional metering charges (VAGO 2015, 29). The metering bill payment was a flat charge, not adjusted according to household income. It was anticipated that customers would make equivalent or larger savings through reduction in bills because the meters allow more detailed feedback on electricity use, and facilitate use of new flexible pricing tariffs with cheaper consumption at particular times of day (called "time of use" tariffs). The AMI Program officially finished at the end of 2013, and a rebate was offered to customers if smart meter installation had still not been attempted at their property by the end of June 2014 (VAGO 2015, 29).

4.1.1. The promise

The policy discourse of the AMI Program at its outset mostly concerned economic issues to do with the privatisation of the Victorian electricity market and the foundation of the Australian National Electricity Market (NEM), which both happened in the period preceding agreement on the AMI Program (1998 marked the start of the NEM, and privatisation of Victoria's electricity sector occurred in the mid-to-late 1990s). Business involvement in the AMI Program, and business priorities, were therefore high from the outset. It was decided that customers in the electricity market (not just large businesses but also smaller customers – small businesses and householders) needed to have better information about how much electricity they were consuming in order to make informed decisions on new time of use tariffs, designed to improve market function (DPI 2007; ESC 2004). Real-time information and feedback on electricity consumption was not possible with traditional mechanical meters, as these only give information on total consumption, not time delimited. The main advantage of the AMI Program was thus framed as improving electricity market function, with the Victorian government stating:

> Customers will benefit from *the enhanced competition in the retail electricity market* associated with the timely and efficient rollout of AMI. (DPI 2007, 16, emphasis added)

Furthermore, the Victorian Essential Services Commission (the state's utility regulator) identified the three main benefits of the new digital metering as follows:

> – Provide the capacity and incentive for customers to manage their electricity consumption more efficiently.
>
> – Increase retail price efficiency.
>
> – Provide distributors with the capability and incentive to introduce more efficient pricing to retailers. (ESC 2004, 10–11)

There was thus a strong emphasis on economic efficiency and benefits to business, and very little was promised at the outset – in the planning stages of the AMI (2004–2008) – regarding social and environmental objectives.

4.1.2. Post-implementation

It was not long after the start of the AMI Program implementation in 2009 that problems started to emerge. Tensions emerged in particular around how the costs of the AMI Program were being distributed, and the social equity issues arising from this (VAGO 2009; Victorian Energy Minister Michael O'Brien 2011). The AMI Program was structured in such a way that financial risks were mostly borne by customers rather than the utilities or government. Moreover, additional costs were ascribed to all households from January 2010, regardless of whether or not they already had a new meter installed (VAGO 2015). Furthermore, as noted, these costs were charged to households at a flat rate, with no discount for low-income households. Alongside these specific social equity concerns, there were a number of broader governance issues raised about the AMI Program, including: the degree of public sector oversight, the exclusion of retailers from decision-making, and data privacy (see Deloitte 2011, 9; VAGO 2015, ix). Such were the concerns about the AMI Program that the Victorian Auditor General (VAGO) investigated it twice, first in 2009 and then 2015. VAGO was highly critical, for example, commenting in its second report that:

> The reality of the smart meter rollout is that the State approved a program, many of the costs of which it could not directly control, nor drive many of the benefits ascribed to it … Despite improvements to customer education … market research conducted in early 2014 found that two-thirds of Victorians did not understand what the benefits provided through smart meters are, and many are still unaware of their ability to help minimise energy bills. (Victorian Auditor General, cited in VAGO 2015, viii)

Public protests and campaign groups emerged such as *Stop Smart Meters Australia* and a dedicated anti-AMI Program political party *People Power Victoria* in the 2010 state elections, with central campaigning issues including negative health effects because of radio frequency emissions from wireless digital meters, rising bills, and privacy concerns (People Power Victoria 2016; SSMA 2015).

Thus, despite initial optimism around the State of Victoria providing a positive "best practice" demonstration of a new electricity metering policy for the rest of Australia to follow (see, for example, Marchment Hill Consulting 2009; NSMP 2008, 4), the AMI Program began to be viewed as policy failure (Lovell 2017), and from 2013 onwards there was a flurry of Australian federal and state government documents explicitly stating that the AMI Program would not be replicated elsewhere (Department of State Growth 2015; NSW Minister for Resources and Energy 2014b; Queensland Department of Energy and Water Supply 2013).

There are a number of ways in which the new technology that was the centerpiece of the AMI Program – the digital meters and their associated communication systems – contributed to these problems, particularly with regard to rising costs. In the mid-to-late 2000s, the period in which the AMI Program was being planned and operationalised, the AMI digital metering and communications technology was very new. The State of Victoria was one of the first large-scale mandatory digital metering programmes world-wide, such that the meters had to be specially ordered for the AMI Program – there was no existing "off the shelf" option, as an interviewee explained:

> Victoria's [metering] specification was one of the leading specifications in the world and it got used by other countries as foundational inputs to their own decisions in relation to smart metering rollouts. So there was a definite consciousness that … you could not go to a metering provider and just pick from the shelf of a range of meters and say 'I'll have that one with those functions and that communications technology please' It just didn't exist. (Interview, Victorian Government employee, November 2016); and

> The [Victorian] government took leadership in developing the functionality specifications of smart metering. It was a global first. (Interview, Victorian Utility manager, November 2016)

One effect of this use of new technology was the rising costs of the AMI Program, as manufacture, implementation and "bug-fixing" all took longer than anticipated, as is reasonably typical for new technologies being implemented on a large-scale (Flyvbjerg, Skamris Holm, and Buhl 2003). Thus, for example, in a 2011 AMI Program review report, commissioned by the new Victorian government in the aftermath of the 2010 state elections, the slow pace of two-way communications on the meters was noted:

> Of the AMI meters installed to date, less than 30% are live and communicating with the distributors. (Deloitte 2011, 17); and

> To date, very few meters have been installed in the most difficult areas of Victoria to serve (where physical barriers such as mountains and long distances between meters and collectors cause problems for the communications network). In most of these difficult areas, the wireless communications networks have not yet been established. (Deloitte 2011, 106)

Thus one of the especially problematic technical issues was the installation of communication systems in rural areas of Victoria, as most experience at the time was based on trials in much denser urban locales. Victoria's AMI Program was a mandatory state-wide programme, so it was not possible to simply leave out customers that proved difficult to connect. The communication system installed by one of the five Victorian distribution companies – SP Ausnet, using a WiMAX technology – never functioned well enough in rural areas of the State, and eventually had to be dropped, with additional costs met by SP Ausnet, despite appeals to the Regulator for customers to be charged with these costs (AER 2013; Coyne 2014).

A second effect from use of novel smart metering technologies was a revision downwards in the expected benefits for customers. These benefits were recognised to have been overstated in the original cost-benefit analysis completed for the AMI Program, leading the Victorian Auditor General (VAGO) to conclude:

> Consumers have realised only a few direct benefits from smart meters … with only 90 per cent of expected benefits realised to date. (VAGO 2015, 28)

As VAGO highlighted, it was households (and small businesses) who missed out most on financial benefits, compared with what was predicted, with only a very small proportion of benefits realised for customers, and most benefits flowing to the utilities (Table 1).

Social equity objectives were, as noted, not a feature of the AMI Program at its outset, but did become more central over time as implementation was underway and there was increasing recognition of the social equity implications of the AMI Program. For example, in response to the 2011 AMI Program government review, the Consumer Utilities Advocacy Centre (CUAC) called for equity issues to be considered in more depth:

> We have taken a position that the smart meter roll out should continue, albeit in a modified form so that net benefits to consumers are maximised and equity issues are addressed. (CUAC 2011, 51)

One outcome of the government's 2011 AMI Program review was that time of use tariffs were not allowed to be implemented at the same time as the new digital meters, because of concerns about rising costs for households. Most consumer group bodies in Victoria supported this suspension in time of use tariffs, for example, as the Victorian Council of Social Service (VCOSS) explained:

> While recognising that smart meters and associated tariff structures, such as time of use pricing, can benefit some households, VCOSS is concerned that there may be other households which can't shift their electricity

Table 1. Victorian AMI program benefits realised (2014) against cost-benefit analysis predictions (2011).

Benefit category	2011 CBA forecast benefit 2008–2014	Actual benefit realised 2008–2014	Percentage of 2008–2014 forecast realised
Avoided cost of accumulation meters	$579.31	$591.99	102.19
Network operational efficiency	$218.94	$107.98	49.32
Tariffs, products and demand management	$9.19	$0.23	2.50
Total	$807.44	$700.20	86.72

Source: amended from VAGO 2015 Figure 3A, p. 32.

consumption without a loss of comfort or wellbeing, such as older people, those with a disability or chronic illness, the unemployed or those at home with small children. These households may face disproportionate price increases, exacerbating financial hardship. (VCOSS 2010, 1)

However, with time of use tariffs disallowed, much of the forecast household economic benefits – promised during the AMI Program planning stages – were no longer possible. Tensions with the business community arose, with many unhappy with the change in policy direction, as a director of one of Victorian distribution utilities described:

the challenges are … these programs are expensive, and as a Board you require certainty that the rules of recovery are not going to change. I mean the same way if you're building a road or whatever. In a private enterprise you need some certainty that the mechanism you use to invest initially is going to sustain through your recovery, and they're not going to try and change it halfway through. So that is a challenge. (Interview, December 2016)

The Victorian government also was in a position where it became impossible to fulfil its original targets for household economic benefits, as one interviewee commented:

The benefits from time of use pricing have been a lot less than anticipated because there is still this fear of time of use pricing. Still there's concern that it's there just to rip people off. And [time of use pricing] was very much a large part of why AMI was supported in the first place. (Interview, Victorian Government employee, November 2016)

More broadly, in response to criticism by VAGO, the government stressed the need to take a longer-term perspective to evaluate the benefits of the AMI Program, including capturing unexpected benefits, such as improved network voltage information:

smart metering provides information, not just about consumption, but also about power quality, which is essential to the distribution businesses. It is really valuable. In fact, the distribution businesses have found better benefits in this regard than were anticipated …. it's a big plus to them … It's probably one of those areas where if you were still running a program, you would go back and say, right, now let's quantify this. Let's fully evaluate this. Does this displace other benefits that we're not achieving? (Interview, Victorian Government employee, November 2016)

In other words, because of its pioneering nature, not all benefits and costs of the AMI Program were captured in the several cost-benefit analyses that were completed: unexpected costs and benefits came to light only once the new technologies were implemented. For example, one important social equity cost – that only became apparent once implementation was well underway – was the ease at which electricity disconnections can take place for customers who are behind with payments. With digital meters the disconnection can be done remotely, and is thus much easier (and cheaper) for the utility:

… in Victoria there appears to be a strong link between the roll out of smart meters (which enable disconnections to be done remotely instead of on a site visit) and an increase in the disconnection completion rate, and therefore also an increase in households being disconnected multiple times over a three-year period. (St Vincent de Paul Society and Alviss Consulting 2016, 5)

This represents a financial benefit for utilities but comes at a significant social equity cost for low-income households. Again, we see here the tension between business objectives and social equity, with these tensions only becoming apparent once implementation was underway.

In summary, the AMI Program digital meters and associated communications technology were very new at the time of planning and implementation, and this is part of the reason why the promise of the AMI Program – centred on economic efficiency objectives – was not delivered according to plan. It is an instance of the technologies of government not being stabilised at the outset, as well as the social aspects of the programme not being given adequate attention in the Program's early stages. The empirical evidence outlined above demonstrates how the smart grid technology – the digital meters and associated communication systems – had a strong influence on the government rationality, with changes being made in response to implementation of the new digital metering as unexpected difficulties arose, which acted not only to disrupt the policy promise of the AMI

Program, but more profoundly, and more positively, to encourage greater consideration of social equity issues within the programme.

4.2. SGSC

SGSC was funded by the Australian federal government in 2009, around the same time as the start of the AMI Program, with AUS$100 million provided by government, along with a further $390 million contributed by industry (directly and in-kind) in order to undertake " ... a commercial-scale trial deployment" (AEFI 2014a, 9) of smart grids in Australia. The funding was announced in the May 2009 federal budget, and consortiums of energy utilities and other energy sector organisations (e.g. universities, industry bodies, local government) were given until January 2010 to prepare and submit proposals (DEWHA 2009a, 12). The core purpose of SGSC was stated at the outset as

> ... aimed at creating, in one Australian city, town or region, an energy network that integrates a smart grid with smart meters in homes, to enable greater energy efficiency, reduced emissions and the use of alternative energy sources, such as solar power. (Commonwealth of Australia 2009, 199)

It was expected at the time that SGSC would be completed by September 2013, but the customer applications trial was later extended to February 2014 (ANAO 2014, 14), with the final SGSC reports published in June and July 2014 (AEFI, 2014a, 2014b; Langham et al. 2014).

In mid-2010, the SGSC funding was awarded to a consortium led by the utility EnergyAustralia (later renamed Ausgrid), along with several project partners including IBM Australia, CSIRO, and the City of Newcastle. From 2010 to 2014, SGSC trials took place in the state of New South Wales, predominately in the cities of Newcastle and Sydney. SGSC comprised five largely separate "work-streams" of smart grid investigation and testing on: information and communication technology platforms, grid applications, customer applications, distributed generation and storage, and electric vehicles (AEFI 2014a, 13). Thus, unlike the Victorian AMI, SGSC was not just about digital meters, but involved testing a whole range of smart grid technologies, including electric vehicles, batteries, grid sensors and wind turbines. However, although large in scope (overall c17,000 households were involved in the trials), it was much smaller scale than the Victorian AMI Program (which installed new metering in over 2 million households, state-wide).

Furthermore, SGSC was purposefully set up as an experiment, to facilitate learning about smart grids. All data and reports from SGSC were made publically available, initially hosted by Ausgrid on an online "Information Clearing House", and later by the Australia government through their central data repository "data.gov.au". Several interim reports were written for each trial, describing the work in the progress. There were also a number of final reports published, including a 668 page summary "National cost benefit assessment" report (AEFI 2014b), as well as more specialist reports, such as on customer research (Langham et al. 2014).

4.2.1. The promise
The promise of SGSC covered a number of different policy goals, including social and environmental ones. In the government's pre-deployment report, seven core objectives for *Smart Grid, Smart City* included: optimising societal benefits by prioritising applications; building awareness of the benefits and obtaining buy-in from industry and customers; and developing an innovative solution that can serve as a global reference case (DEWHA 2009b, 14). Notably, there was close attention to climate change, with a policy promise of a minimum of 3.5 megatonnes of CO_2-e (carbon dioxide equivalent) reductions per annum (DEWHA 2009b, 34). Indeed, the SGSC proposal was developed and funded from the federal government's National Energy Efficiency Initiative, and this in part accounts for its early emphasis on energy efficiency and climate change:

> The new *National Energy Efficiency Initiative: Smart Grid, Smart City* will use twenty-first century technology to assist Australia's transition to a low carbon economy by encouraging a smarter and more efficient electricity network. (DEWHA 2009a, 3); and

Consumer applications enabled by smart meters can assist users to understand and control their energy use, reducing peak loads and delivering greenhouse gas savings. (DEWHA 2009a, 3)

Societal benefits were also considered, but the stated objectives were relatively weak, including issues such as increasing network reliability, and slowing growth in energy prices (DEWHA 2009b, 35).

4.2.2. Post-implementation

There are a number of ways in which the promise of SGSC changed as implementation took place. Already by the time the federal government and Ausgrid signed the funding agreement in 2010 the programme objectives had shifted slightly, with no direct mention of societal benefits, though "environmental benefits" were still included, with one objective to:

Build public and corporate awareness of the economic and *environmental benefits* of smart grids and obtain buy-in from industry and customers. (Funding Agreement between the Commonwealth of Australia and Ausgrid, Schedule 1, clause 1.5; cited in ANAO 2014, 101, emphasis added)

The Australian National Audit Office (ANAO) reviewed SGSC in early 2014, in the final stages of the programme, and acknowledged a range of problems arising during implementation, commenting:

The $100 million Smart Grid, Smart City demonstration program was established to implement or trial a range of new technologies in a challenging environment. These challenges included technological issues, consumer resistance to smart metering technologies, regulatory reform in the electricity sector, and responsibility for the program being transferred across four [government] departments between 2009 and 2013. (ANAO 2014, 16)

The ANAO was most critical of the retail or "customer applications" trial – the one which most directly involved households, and which cost AUS$20million. In Ausgrid's original proposal to government for the SGSC funding the utility promised " ... to include up to 50,000 customers in the customer applications trials – 30,000 with 'mandatory' smart meter installations and provision of feedback technologies as part of a 'network' trial, and up to 20,000 in the 'opt-in' retail trial" (ANAO 2014, 126). The federal government in its 2009 pre-deployment report had indicated that around 9000–10,000 customers would allow for robust data (DEWHA 2009b, 10), so Ausgrid's plans were well in excess of that. However, not long after SGSC implementation started, Ausgrid put forward a new recommendation of only 4453 customers in the retail trial " ... with a 'stretch target' of 8333" (ANAO 2014, 126). In reality Ausgrid did not even obtain those numbers of households, with only around 4000 customers actually completing the trials: 8508 customers signed up, but a large number – 4508 – " ... either opted out of the trial or were removed by Ausgrid due to technical or installation issues" (ANAO 2014, 104), leading the ANAO to conclude:

As a consequence of the reduced timeframe and participation rate, the retail trial has not generated the volume or breadth of data that was initially envisaged. (ANAO 2014, 104)

Within the customer applications trial there were a range of unanticipated issues in recruiting households, as well as in installing the new digital meters and enabling the wireless communications, as the final SGSC project report explains:

In the SGSC Program, almost 30 per cent of sites were found to be unsuitable for the deployment of the smart meter infrastructure ... despite a program of pre-qualifications and site visits. (AEFI 2014b, 104, emphasis added)

Somewhat mundane issues, such as the larger size of the new digital meters, created problems which were not foreseen, as described by the project partners:

The size of the smart meter board created installation challenges in many cases during the trial. The larger footprint of the smart meters in comparison with non-smart meters (and even other smart meters) meant that, in many cases it was unable to be installed due to insufficient space .... (AEFI 2014b, 105);

So too did difficulties in ordering high volumes of digital meters, and with the communication network that the meters use to send and receive data:

Ausgrid ... experienced a number of technical issues in deploying the smart meters that were required for the retail and network trials. These included problems sourcing appropriate meters, with some failing initial software testing, and ... in relation to the signal strength required for the smart meters to communicate data. As a result, the number of suitable households was significantly reduced. (ANAO 2014, 103)

The high drop out rate of customers signing up to the retail trial and then not continuing (a 50% drop out rate) was " ... much higher than expected" and interestingly is attributed in part to " ... negative publicity regarding other smart meter rollouts (particularly in Victoria)" (ANAO 2014, 103); thus demonstrating how the two smart grid projects interacted and influenced each other during their implementation.

In relation to its economic objectives, the financial benefits actually arising from SGSC were a lot lower than expected, as was experienced with the Victorian AMI Program. The cost-benefit analysis completed for the SGSC final report in 2014 found that instead of $5billion predicted annual benefits from smart grid implementation (if rolled out nationally, as per 2009 federal government business case projections (DEWHA 2009b, 7)), there were judged to be only $887 million annual benefits in practice (AEFI 2014b, 235). Costs were also higher than expected. For instance, despite the far fewer numbers of households involved in the customer applications trial (20% of the original number promised), and the trials running only for one year rather than two, the cost was the same, at a total cost of $20million – equating to an average of $5000 per customer (ANAO 2014, 25).

In terms of environmental objectives, specifically in relation to climate change, the overall finding of SGSC was that greenhouse gas emissions rise with a smart grid. As mentioned, the original 2009 federal government target for annual emission reductions was 3.5 megatonnes (Mt) of CO_2-e (DEWHA 2009b). However, although under the low economic scenario modelled in the cost-benefit analysis the average annualised difference in emissions under the smart grid case was 2.15 Mt lower each year, under the medium and high economic cases smart grid annual emissions intensity was higher, at 6.95 Mt and 13.71 Mt, respectively (AEFI 2014b, 236). It is not clearly explained in the SGSC reports why there was found to be a rise in greenhouse gas emissions, but it appears to be related to the economic scenarios used for the cost-benefit analysis – essentially with lower growth there are fewer carbon emissions. In other words, the unexpected rise does not stem directly from SGSC trial data, but rather from the macroeconomic scenarios used to underpin the cost-benefit analysis (Energeia 2014, 147). But, perhaps as significantly there is evidence of waning stakeholder interest in climate change by the time SGSC is drawing to a close in 2013–2014. For example, in a 2013 SGSC Stakeholder Engagement Report it was noted that:

Greenhouse gas and other emissions were of lower interest [amongst stakeholders] as these do not directly fall within the [electricity] network remit. (Langham et al. 2013, 29)

In summary, SGSC changed as it was implemented: objectives shifted, fewer households participated than was originally planned, costs rose and benefits declined. In contrast to the Victorian AMI Program – where equity considerations became a greater component of the programme over time – with SGSC equity declined in importance and economic objectives became more prominent. In both cases, there is evidence that smart grid technologies did not act in ways that fulfilled the original discursive policy promise of smart grids.

5. Summary and conclusions

In summary, this paper has examined what happened in practice with two large smart grid programmes in Australia: SGSC and the Victorian AMI Program. It has been demonstrated that the promise of these two smart grid programmes has not for the most part been realised: unforeseen technical and societal difficulties occurred, costs escalated and many of the planned benefits did not come to pass. This resulted in a policy retreat from smart grids in Australia, with, for example, the industry group Smart Grids Australia commenting:

Leadership and the role of government in the development of a national [smart grid] roadmap ... [is] unclear, ad hoc and piecemeal. (Smart Grid Australia 2014, 5) and more specifically the NSW state government – along with other Australian States – rejecting the type of smart metering programme that was implemented in Victoria:

> Not only were Victorian customers not given a choice of meters, they were also charged the upfront cost of the meter and its installation, a decision which is still costing them. *The [NSW] Government has listened to customers and that is why ultimately customers will decide what they want and when they want it.* (NSW Minister for Resources and Energy 2014a, emphasis added)

The focus of analysis of the papers in this Special Issue is on social justice, sustainability and equity issues, and the two cases examined here are instructive in demonstrating the shifting tensions between "smart" and "sustainability". The Victorian AMI Program was out the outset about economic objectives, and there was very little promised with regard to social and environmental benefits. However, as implementation occurred this did change, and social equity issues were more integrated into the Program. With SGSC there was the opposite movement: environmental and social benefits were prioritised at the start, in the early discourse around SGSC, with climate change in particular having a strong emphasis. However, by the end of SGSC, there was much less focus on social and environmental outcomes. The final reporting on SGSC was centred on a national cost-benefit analysis, with economic objectives having clear priority. Furthermore, GHG emissions were shown to have actually risen with smart grids, according to the SGSC national cost-benefit modelling results.

A governmentality approach allows us to better understand the changing relationship between policy discourse and practice with regard to smart grids. In the two cases examined here, technology failures and set backs in the process of smart grid implementation acted in ways that changed the previously optimistic rationality around smart grids in Australia. In other words, it is a case where smart grid technologies have not only failed to support, but have actually undermined, the original promise of smart grids, and hence it is a case of smart grid technologies challenging rather than accomplishing governance. The tensions observed here between "smartness" and social and environ-mental equity, are not new (Gabrys 2014; Hollands 2008; Klauser, Paasche, and Söderström 2014), but, using the implementation of technologies as the focus of empirical analysis allows for different insights – with everyday mundane difficulties linked to, and born out of, more fundamental tensions. It is these everyday mundane difficulties in the process and practice of governance that governmen-tality speaks to so well, and, where there is judged to be scope for greater consideration of the mul-tiple ways in which smart grid technologies have undermined (or supported) smart grid policy initiatives. The most valuable insight provided by a governmentality lens for those interested in pro-cesses of technological change concerns the close two-way relationship between the rationalities and technologies of government. What this contribution adds is two detailed smart grid/energy empirical cases that demonstrate the agency of governmental technologies to disrupt and alter rationalities; with a myriad of small relatively commonplace technical issues of policy implementation collectively having a profound effect on the impact and direction of smart grids policy in Australia.

Disclosure statement

No potential conflict of interest was reported by the author.

Funding

This work was supported by the Australian Research Council under Grant number FT140100646.

References

AEFI. 2014a. *Smart Grid, Smart City: Shaping Australia's Energy Future – Executive Report*. Canberra: Arup, Energeia, Frontier Economics, and the Institute for Sustainable Futures (University of Technology Sydney) (AEFI).

AEFI. 2014b. *Smart Grid, Smart City: Shaping Australia's Energy Future – National Cost Benefit Assessment*, 668. Sydney: Report for Ausgrid.

AER. 2013. *Final Decision: Advanced Metering Infrastructure Review SPI Electricity Pty Ltd 2012–2015 Budget and Charges Applications*. Canberra: Australian Energy Regulator.

Akrich, M. 1992. "The De-Scription of Technical Objects." In *Shaping Technology/Building Society: Studies in Sociotechnical Change*, edited by W. E. Bijker and J. Law, 205–224. Cambridge, MA: MIT Press.

ANAO. 2014. *Administration of the Smart Grid, Smart City Program*. Canberra: The Auditor-General, Australian National Audit Office (ANAO).

Backstrand, K., and E. Lovbrand. 2006. "Planting Trees to Mitigate Climate Change: Contested Discourses of Ecological Modernization, Green Governmentality and Civic Environmentalism." *Global Environmental Politics* 6: 50–75.

Bulkeley, H., G. Powells, and S. Bell. 2016. "Smart Grids and the Constitution of Solar Electricity Conduct." *Environment and Planning A* 48: 7–23.

Commonwealth of Australia. 2009. *Budget Measures: Budget Paper No. 2: 2009–2010, p. 199*. Canberra: Commonwealth of Australia Federal Government.

Coyne, A. 2014. "AusNet Adds $175 m to Cost of Vic Smart Meter Rollout." *IT News*. Accessed 13 November 2017, see https://www.itnews.com.au/news/ausnet-adds-175m-to-cost-of-vic-smart-meter-rollout-397790

CRA International, Impaq. 2005. *Advanced Interval Meter Communications Study*. Melbourne: report for the Department of Infrastructure – Energy and Security Division.

CUAC. 2011. *Review of the Victorian AMI Program: Response to Issues Paper*. Melbourne: Consumer Utilities Advocacy Centre (CUAC).

Dean, M. 1999. *Governmentality: Power and Rule in Modern Society*. London: Sage.

de Jong, M., S. Joss, D. Schraven, C. Zhan, and M. Weijnen. 2015. "Sustainable-smart-resilient-low Carbon-eco-Knowledge Cities; Making Sense of a Multitude of Concepts Promoting Sustainable Urbanization." *Journal of Cleaner Production* 109: 25–38.

Deloitte. 2011. *Advanced Metering Infrastructure Cost Benefit Analysis*. Canberra: Department of Treasury and Finance.

Department of State Growth. 2015. "Tasmanian Energy Strategy: Restoring Tasmania's Energy Advantage".

DEWHA. 2009a. *Smart Grid, Smart City Grant Guidelines: The National Energy Efficiency Initiative*. Canberra: Department of the Environment, Water, Heritage and the Arts.

DEWHA. 2009b. *Smart Grid, Smart City: A New Direction for a New Energy era*. Canberra: Department of the Environment, Water, Heritage and the Arts, Commonwealth of Australia.

DPI. 2007. *Victorian Government Rule Change Proposal – Advanced Metering Infrastructure Rollout*. Melbourne: Department of Primary Industries (DPI).

Energeia. 2014. *Modelling Inputs Compendium: SGSC National Cost Benefit Assessment*. Sydney: Energeia, Ausgrid.

ESC. 2004. *Mandatory Rollout of Interval Meters for Electricity Customers*. Melbourne: Essential Services Commission.

Flyvbjerg, B., M. K. Skamris Holm, and S. L. Buhl. 2003. "How Common and how Large Are Cost Overruns in Transport Infrastructure Projects?" *Transport Reviews* 23 (1): 71–88.

Gabrys, J. 2014. "Programming Environments: Environmentality and Citizen Sensing in the Smart City." *Environment and Planning D: Society and Space* 32: 30–48.

Graham, S., and S. Marvin. 2001. *Splintering Urbanism: Networked Infrastructures, Technological Mobilities and the Urban Condition*. London: Routledge.

Higgins, V., and W. Larner. 2010. *Calculating the Social: Standards and the Reconfiguration of Governing*. Basingstoke: Palgrave Macmillan.

Hollands, R. G. 2008. "Will the Real Smart City Please Stand up?" *City* 12: 303–320.

Hughes, T. P. 1983. *Networks of Power: Electrification in Western Society 1880–1930*. MD: The John Hopkins University Press.

Karvonen, A., and B. van Heur. 2014. "Urban Laboratories: Experiments in Reworking Cities." *International Journal of Urban and Regional Research* 38: 379–392.

KEMA. 2013. *National Smart Meter Infrastructure Report*. Canberra: Kema DNV produced for Department of Resources, Energy and Tourism.

Klauser, F., T. Paasche, and O. Söderström. 2014. "Michel Foucault and the Smart City: Power Dynamics Inherent in Contemporary Governing Through Code." *Environment and Planning D: Society and Space* 32: 869–885.

Langham, E., J. Downes, T. Brennan, J. Fyfe, S. Mohr, P. Rickwood, and S. White. 2014. *Smart Grid, Smart City, Customer Research Report. Report Prepared June 2014*. Sydney: By the Institute for Sustainable Futures as Part of the AEFI Consortium for Ausgrid and EnergyAustralia.

Langham, E., N. Ison, T. Brennan, J. Downes, L. Boronyak, and S. White. 2013. *Smart Grid, Smart City: Analysis and Reporting. Stakeholder Engagement Report*. Sydney: Institute for Sustainable Futures, UTS for Ausgrid.

Lansing, D. M. 2011. "Realizing Carbon's Value: Discourse and Calculation in the Production of Carbon Forestry Offsets in Costa Rica." *Antipode* 43: 731–753.

Lazar, J., and M. McKenzie. 2012. *Australian Standards for Smart Grids – Standards Roadmap*. Canberra: A report for the Federal Department of Resources, Energy and Tourism.

Li, T. M. 2007. *The Will to Improve: Governmentality, Development, and the Practice of Politics*. Durham: Duke University Press.

Lovell, H. 2017. "Are Policy Failures Mobile? An Investigation of the Advanced Metering Infrastructure Program in the State of Victoria, Australia." *Environment and Planning A* 49: 314–331.

Luque-Ayala, A., and S. Marvin. 2016. "The Maintenance of Urban Circulation: an Operational Logic of Infrastructural Control." *Environment and Planning D: Society and Space* 34: 191–208.

Marchment Hill Consulting. 2009. "Victorian AMI Program – Presentation to the National Smart Metering Program." Accessed 21 January. https://link.aemo.com.au/sites/wcl/smartmetering/Documentlibrary/Forms/DispForm.aspx?ID=156.

McGuirk, P., H. Bulkeley, and R. Dowling. 2014. "Practices, Programs and Projects of Urban Carbon Governance: Perspectives from the Australian City." *Geoforum* 52: 137–147.

Mitchell, C. 2008. *The Political Economy of Sustainable Energy*. Basingstoke: Palgrave Macmillan.

Morgan, K. 2004. "The Exaggerated Death of Geography: Learning, Proximity and Territorial Innovation Systems." *Journal of Economic Geography* 4: 3–21.

NSMP. 2008. *National Smart Metering Program Pilots and Trials 2008 Status Report to the Ministerial Council on Energy*. Canberra: National Smart Metering Program (NSMP).

NSW Minister for Resources and Energy. 2014a. "Media Release: NSW Gets Smart about Smart Meters".

NSW Minister for Resources and Energy. 2014b. "NSW Gets Smart about Meters".

Nye, D. E. 1992. *Electrifying America: Social Meanings of a new Technology, 1880–1940*. Cambridge, MA: The MIT Press.

People Power Victoria. 2016. "People Power Victoria – Homepage." http://www.peoplepowervictoria.org.au/home.

Power, M. 1994. "The Audit Society." In *Accounting as Social and Institutional Practice*, edited by A. G. Hopwood and P. Miller, 299–316. Cambridge: Cambridge University Press.

Queensland Department of Energy and Water Supply. 2013. *The 30-year Electricity Strategy. Discussion Paper – Powering Queensland's Future*. Brisbane: Queensland Government – Department of Energy and Water Supply.

Smart Grid Australia. 2014. *Towards Australia's Energy Future: 2014 Update*. Canberra: Smart Grid Australia (SGA).

SSMA. 2015. "Stop Smart Meters Australia-Fighting for Your Financial & Physical Health, Privacy, and Safety in Australia." http://stopsmartmeters.com.au/.

St Vincent de Paul Society, Alviss Consulting. 2016. *Households in the Dark: Mapping Electricity Disconnections in South Australia, Victoria, New South Wales and South East Queensland*. Melbourne: St Vincent de Paul Society and Alviss Consulting.

Turnheim, B., and F. W. Geels. 2012. "Regime Destabilisation as the Flipside of Energy Transitions: Lessons from the History of the British Coal Industry (1913–1997)." *Energy Policy* 50: 35–49.

Unruh, G. C. 2002. "Escaping Carbon Lock-in." *Energy Policy* 30: 317–325.

VAGO. 2009. *Towards a "Smart Grid" – The Roll-out of Advanced Metering Infrastructure*. Melbourne: Victorian Auditor-General.

VAGO. 2015. *Realising the Benefits of Smart Meters*. Melbourne: Victorian Auditor-General's Office (VAGO).

VCOSS. 2010. *VCOSS Briefing Paper – Smart Meters/Advanced Metering Infrastructure (AMI)*. Melbourne: Victorian Council of Social Services (VCOSS).

Victorian Energy Minister Michael O'Brien. 2011. "Smart Meters here to Stay Despite Cost Blow-out." http://www.abc.net.au/news/2011-12-14/smart-meter-roll-out-continues-despite-cost-blow-out/3730522.

Viitanen, J., and R. Kingston. 2014. "Smart Cities and Green Growth: Outsourcing Democratic and Environmental Resilience to the Global Technology Sector." *Environment and Planning A* 46: 803–819.

Winner, L. 1977. *Autonomous Technology: Technics-out-of-Control as a Theme in Political Thought*. Cambridge: MIT Press.

Zimmerman, A. D. 1995. "Toward a More Democratic Ethic of Technological Governance." *Science, Technology, & Human Values* 20: 86–107.

Smart meter data and equitable energy transitions – can cities play a role?

Jess Britton ⓘ

ABSTRACT
The ability of smart cities to address multiple priorities is currently receiving a great deal of attention from governments, industry and academia; however, there are relatively few explorations of how smart cities will be enacted and the equity implications of various elements of the agenda. This paper aims to begin to address this by considering the case of smart meter data in Great Britain and reviewing the potential for public interest benefits to be realised at both a national and city-scale. In order to interrogate how societal benefits of smart metering might be enabled or constrained at different scales, both transition management approaches to understanding the intricacies of transforming socio-technical systems, and urban governance literatures which explore the problems in aligning "smart" and "sustainable" agendas, are engaged with. The case suggests that there is great potential for smart meter data to deliver public interest benefits and in doing so contribute to equity in energy transitions through promoting the accrual of benefits to collective rather than solely individual or commercial interests. Despite this, under the current smart metering framework city-scale actors are largely excluded from utilising smart metering data unless they partner with a large incumbent company. This suggests that the configuration of national smart metering arrangements could be further embedding a marketised approach to smart cities and demonstrates the complex interlinking of socio-political processes operating across multiple scales in energy transitions.

1. Introduction

References to "smartness" and "smart systems" have exploded in public policy, academic and industry arenas in recent decades with great expectations attributed to the ability of new technologies (particularly IT) to address modern societal challenges. The difficulty in precisely defining what is meant by "smart" is also well established (Luque *et al.* 2014, Kitchin 2015), and this paper considers in detail the intersection of two aspects of the "smart" agenda: smart meter data and smart cities. A particular focus is given to how these two agenda intersect in terms of delivering public interest benefits.

Smart electricity and gas meters are currently being rolled out across many countries[1] in order to support the development of smart energy systems that both provide consumers with more accurate billing and consumption information and create opportunities for new energy efficiency services, demand side response (DSR) and storage (DECC 2015a). As such, smart meters are expected to be one of the "critical building blocks" in the development of a low-carbon economy (DECC 2015b, p. 4).

The UK Government has set up a programme to ensure domestic and small non-domestic custo-mers in Great Britain (GB) are provided with smart meters by the end of 2020 and these smart elec-tricity and gas meters will provide an unprecedented volume and granularity of energy consumption data (DECC 2015b). Although there has been extensive discussion of the costs, benefits and privacy requirements of smart meters and of the potential for smart meter data to change consumer beha-viours, the debate has largely been focused on using data in consumer feedback, enabling the future smart grid, and for commercial applications. Very limited attention has been paid to the potential for smart meter data to be used to serve the public interest, for example to improve public policy-making or to be used by socially and/or environmentally motivated groups such as community energy and fuel poverty organisations.

Given that all energy consumers are paying for the smart meter roll out through their bills, it is important that the broader public interest opportunities of smart meter data are maximised in order to ensure that any benefits of smart meter data are as inclusive as possible, and not limited to only the most proactive of consumers. This is particularly the case as it is unlikely that all consumers will engage equally with their smart meter data in order to realise direct benefits. Additionally, his-torically, many uses of energy consumption data have been enacted at a city scale through locally based energy efficiency, affordable warmth and behaviour change programmes. It is, therefore, important both to understand the potential for the public interest to benefit from smart meter data and to consider how city-scale uses of energy consumption data might be impacted by the smart metering framework. This paper, therefore, examines the potential for smart meter data to con-tribute to the public interest through a detailed review of the current arrangements in GB and con-siders whether cities have a role to play in realising these public interest benefits.

The paper is based on research carried out as part of a TEDDINET[2] research challenge which ana-lysed the public interest opportunities smart meter data may present at both a national and sub-national scale. This involved a review of policy and academic literatures relating to smart meter data and the public interest, analysis of existing and potential public interest uses of smart meter data at a range of scales and evaluation of the current arrangements for data use for public interest aims.

Already a range of actors, across the public, private and social movement spectrum, are exploring how they could utilise smart meter data. However, it is likely that these different actors and uses will have different implications for the public interest. This paper concludes that public interest uses of smart meter data have been under explored to date and identifies a number of benefits and chal-lenges in addressing this issue. In particular, it is evident that cities and other sub-national actors could play an important role in realising some public interest benefits. These city-scale public interest uses of data include improving the targeting and monitoring of local energy programmes, supporting social housing and health services, providing new opportunities for community energy and enabling local approaches to balancing supply and demand. There is a great deal of debate regarding the potential for new actors and interests to engage in the smart energy system; however, city-scale approaches to smart meter data are likely to encounter particular barriers, demonstrating the diffi-culty for non-incumbent, and particularly local scale, interests to engage in this agenda.

Current approaches to smart meter data in GB are dominated by a focus on (large scale) commer-cial applications of data or on individual benefits to consumers, largely driven by the dominance of incumbents in smart energy governance and also by marketised approaches to smart cities. These approaches risk failing to ensure that the benefits of smart meter data are realised in a socially just manner as collective benefits are marginalised.

Although the focus of this paper is on the specifics of the smart metering framework in GB, the conclusions are also relevant in relation to the social and political shaping of smart energy systems across multiple scales. This includes a number of conclusions relating to the socio-technical transitions literature. In particular, the development of the smart metering agenda has been domi-nated to date by incumbent actors who have promoted discourses of individual consumer gain and commercial data uses, rather than public interest potential. This has resulted in a smart metering framework that largely excludes public interest actors, particularly at the city scale and it is likely that such uses will need to partner with an incumbent energy system actor to realise public interest benefits. This suggests that the configuration of national smart metering arrangements could be further embedding a marketised approach to smart cities and demonstrates the multifaceted inter-linking of socio-political processes operating across multiple scales in complex transitions such as the energy transition.

2. Exploring smart transitions and cities

2.1. Smart-sustainable city discourses

Debates regarding urban futures often refer to "smart-sustainable" urban development, but critical analysis is increasingly recognising that the ideas underpinning both the "smart" and "sustainable" agendas are highly contested with multiple different interests framing the two issues, often with con-flicting values (Viitanen and Kingston 2014, Kitchin 2015, De Jong *et al.* 2015).

The term "smart city" is widely used in policy and academic literature and is generally taken to mean using new technologies (mainly information and communication technologies) and data to improve service delivery and address various economic, social and environmental challenges at a city scale (Centre for Cities 2014). However, underlying this relatively simple definition is a wide range of ideas regarding exactly what constitutes a smart city with multiple smart city definitions pro-moted by various actors (Hollands 2008, Glasmeier and Christopherson 2015).

Despite the contested nature of smart city definitions, the majority suggest that ITs have an impor-tant role to play in addressing urban sustainability problems (Viitanen and Kingston 2014, Carvalho 2015) and tend to unproblematically bring together discourses of smartness and sustainability as complementary and mutually reinforcing (Viitanen and Kingston 2014). Specifically, a number of authors have recognised that smart city discourses often reframe urban sustainability challenges as opportunities for private firms to market products and services (such as smart meters) (Viitanen and Kingston 2014). This reinforces a "green growth" approach that promotes neoliberal, entrepre-neurial forms of sustainable urban development (Hollands 2008) and lacks analysis of the potential conflict between growth-focussed "smart cities" and environmental sustainability. This neoliberal bias of smart city agendas also suggest potential for the focus to be on individual benefits at the expense of broader societal gains and both Glasmeier and Christopherson (2015, p. 6) and Carvalho (2015) highlight that smart city framings tend to lack "data applications that could drive collective rather than individual solutions", with little focus on social equity concerns. Additionally, smart city visions often suggest that smartness will benefit all residents but in reality it is likely that not all resi-dents will possess the economic resources, skills or motivation to access benefits equally. Hollands (2008, p. 315) suggests that this has the potential to create a "two speed city" where the smart agenda materially benefits affluent residents but overlooks a growing precariat.

2.2. Socio-technical urban governance

Drawing on the science and technology studies literature, several authors (such as Hodson and Marvin 2009, Goodspeed 2015) have highlighted that a smart city is defined not just by IT and associ-ated infrastructure but also by the social and organisational structures that facilitate the use of these physical infrastructures. Additionally, looking specifically at smart energy data, we can see that the infrastructure requirements for a smart energy system (smart grids, smart meters, data management systems) are only part of the story with the structuring of systems for data collection, management and use of central importance to how the smart energy agenda will be realised. Seeing the develop-ment of smart city agendas in this socio-technical context allows us to consider both the framing of smart city discourses from an urban governance perspective but also, in relation to smart energy data, to examine how actors at multiple scales may be engaging with one part of the energy transition – smart metering.

Although the socio-technical transitions literature provides a comprehensive framework for analysing change in large complex systems, such as energy, a number of authors have drawn attention to the lack of a sufficient conceptualisation of place and scale in this literature (Hodson and Marvin 2010, Smith et al. 2010, Lawhon and Murphy 2012, Coenen and Truffer 2012). Historically, socio-technical systems have tended to be presented as consisting of nested, largely autonomous, spatial scales (Rutherford and Coutard 2014) with little recognition that the socio-political processes involved in transitions take place a multiple scales or that city networks may be significant.[3] This has resulted in limited consideration of the role of different actors at the city scale, process of negotiation and development, and the interplay between local-scale projects and wider national and international regimes.

As this section outlines analysis of smart city discourses suggests that to date they are dominated by a focus on individual and commercial gain with a lack of exploration of social justice and equality issues. Additionally, the socio-technical transitions literature illustrates the importance of social and organisational structures in shaping physical infrastructure transitions, and highlights how the city-scale is an underexplored element of transition processes. The next section, therefore, analyses the GB smart meter data framework and reviews the potential for public interest benefits to be realised, with the objective of exploring how insights from urban governance literature on smart cities can illuminate the role of cities in socio-technical transitions.

3. The GB domestic smart metering programme

The UK Government has set up a programme to ensure that domestic and small non-domestic customers in GB are provided with smart meters by the end of 2020. The programme aims to install over 50 million smart electricity and gas meters across GB by the end of 2020.[4] The definition of a "smart" meter, although not straightforward, is generally agreed as one which can both "(1) measure and store data at specified intervals and (2) act as a node for two-way communications between supplier and consumer and automated meter management (AMM)" (Darby 2010, p. 445). The data captured by these meters will transform the volume and granularity of energy consumption data available. This is likely to provide scope for dramatic changes in how consumers interact with their energy usage[5] and is central to the development of a smart energy grid that can facilitate greater volumes of low-carbon, flexible generation.

In GB, the precise definition of what constitutes a smart meter is defined in Government policy through the Smart Metering Equipment Technical Specifications (SMETS) (DECC 2014b). The SMETS set out the minimum physical, functional, interface and data requirements of smart gas and electricity meters in GB and are supported by a technical and institutional framework managed via the Smart Energy Code (SEC) and Data Communications Company (DCC) (see Section 3.2).

The Government has forecast that the smart meter roll out will result in economic savings for consumers, with a total net gain of over £6 billion (NPV to 2030) from implementing the smart meter programme. Savings are expected to arise largely through reduced supplier costs (through avoided meter reads, improved debt handling and reduced theft) and reduced energy use from better energy feedback and novel tariff structures (DECC 2014a). Whilst these benefits are important, smart meter data can also benefit the wider public interest through improved public policy-making and the utilisation of data by socially and/or environmentally motivated groups such as community energy and fuel poverty organisations.

The GB smart meter roll out is led by energy suppliers with costs financed by all energy consumers through their energy bills. Despite this, not all consumers are likely to engage directly with their smart meter data and those consumers that do not engage with their data are unlikely to fully experience the direct economic benefits identified in the smart meter impact assessment. It is, therefore, important that the wider public interest opportunities of smart meter data are realised to ensure that any benefits of smart meter data are as inclusive as possible, and not limited to only the most proactive of consumers.

3.1. Smart meter data

Historically, in GB, data on domestic energy consumption have been reliant on large nationally representative surveys and data collected during small research projects and trials. Such data sources (e.g. the Digest of UK Energy Statistics, Annual estimates of domestic and non-domestic electricity and gas consumption, the National Energy Efficiency Data Framework (NEED)) are valuable resources for policy-making and delivery, but their utility is limited due to spatial and temporal aggregation, with the majority of data only available in an annualised format, spatially aggregated or otherwise anonymised (Elam 2016, Britton 2016).

To date, the focus of smart meter discourses has largely been on the potential for smart meter data to benefit individual households (through bills savings and opportunities for innovative tariff structures) and to support the development of a smart energy system. However, in addition to these benefits smart meter data could also inform public policy and other projects motivated by the public interest, indeed many of the current uses of domestic energy consumption data are for public interest uses. These existing uses include national government, local authorities, housing associations, community groups and NGOs using nationally coordinated consumption data and local data on tenure, housing type, deprivation and so on, to support the targeting of energy efficiency programmes and the development of Affordable Warmth Strategies.

3.2. The current smart meter data framework

The smart metering framework puts the energy consumer at the centre of the smart metering programme, giving the consumer control over access and use of data. The SMETS, therefore, require that smart meters are capable of storing half-hourly energy consumption data but that consumers also have the option to view or collect more granular (10-second, near real-time) data via an In-Home display or Consumer Access Device (CAD) connected to the Home Area Network. Only the consumer will be able to see this near real-time data, unless they have authorised another party to access the data (DECC 2012b).

The DCC is responsible for linking consumer's smart meters with energy suppliers, network operators and third parties, as outlined in Figure 1. The terms of the provision of DCC services are set out in

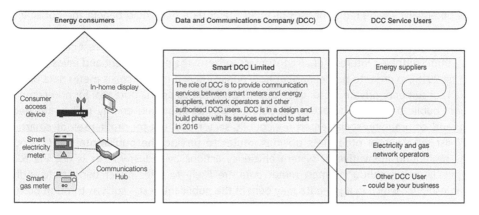

Figure 1. GB smart metering system (DECC 2015b).

the SEC, a new industry code which specifies the technical governance of end-to-end management of smart metering.

Smart meter data access will be tightly controlled with requirements set out in the Data Access and Privacy Framework (DAPF). In addition to access for consumers, the DAPF details the access that suppliers, network operators and third parties can have and establish the purposes for which data can be collected. This includes a requirement for explicit consumer consent for access to daily, half-hourly or more frequent data. At a more detailed level, energy suppliers will be able to access monthly consumption data without customer consent for billing purposes and to fulfil any statutory requirement or licence obligation. Suppliers will also be able to access daily (or less granular) energy consumption data for any purpose except marketing, provided customers have a clear opportunity to opt out. Suppliers must receive explicit (opt-in) consent from the customer in order to access half-hourly energy consumption data, or to use energy consumption data for marketing purposes. A number of exceptions have been specified to this basic framework, for example to allow suppliers to access daily energy consumption data without customer consent where they have reasonable suspicion of theft.

Distribution network operators will only be able to access domestic customers' energy consumption data without consent for regulated purposes and must aggregate and/or anonymise all data, with the process for anonymising data to be approved by Ofgem.

The DAPF also outlines a regime for third-party access to data, referring to a wide range of "third parties" such as energy services companies and switching sites (DECC 2012a). In addition, many public interest users of smart meter data (such as those discussed in Section 3.3) would be considered to be third parties. There are three routes for third parties wishing to access smart meter data:

- Third parties wishing to access data remotely via the DCC must be registered as DCC Users and be SEC signatories. They will be required to verify explicit opt-in consumer consent and provide reminders to consumers about the data that are being collected.
- Alternatively, third parties can enter into a commercial contract with an existing DCC user in order to access data. The third party would be required to gain individual consent to access data and to protect the consumer's privacy.
- Or finally, customers can collect data directly from a CAD[6] and exchange these data with a third party outside of the arrangements in the Privacy and Access Framework. Such arrangements would be governed by a contract between the consumer and third party, and third parties would be bound by relevant legislation such as the Data Protection Act (DECC 2012a).

3.3. Potential public interest uses of smart meter data

There is potentially a wide range of uses of smart meter data that could benefit the public interest. This paper does, however, recognise the difficulty in precisely defining the public interest and identifying when the public interest is served (Rutgers 2015). Whilst the public interest can be defined as the long-term combined interests of all consumers, citizens, the environment and investors (Sustainability First 2015), in order to be clear about how public interest uses of smart meter data differ from individual consumer or private sector uses, this review focusses on the potential for non-commercial, *communal* public benefit to be achieved, defining this as improved energy policy and decision-making, and community, social enterprise, public sector and not-for-profit uses of smart meter data. Whilst the value of other, less obvious, routes to serving the public interest is recognised (such as promoting innovative or system efficiency actions by industry), this is not the focus of this paper. Likewise, although smart meter data are likely to offer direct benefits to individual energy consumers, which in aggregate may benefit the public interest – such as through promoting greater awareness of energy use – these benefits have been well documented and are not the focus of this paper.

In terms of improved public policy, there is great scope for smart meter data to inform the evaluation of existing policy and to aid the development of future policy at both a national and subnational level. For example, the impact of interventions made through schemes such as the Energy Company Obligation could be accurately evaluated. Additionally, measures such as the UK "Low Income High Cost" definition of fuel poverty (Hills 2012), which requires data on energy costs, could be improved with actual data utilised in place of the current reliance on modelled data.

Smart meter data on energy costs and time of use could provide valuable inputs to policy-making and modelling to develop a better targeted policy to address fuel poverty and energy efficiency. This includes replacing the existing annualised energy consumption data with more accurate smart meter data in order to enhance the evidence base for government policy. More broadly, smart meter data could inform government policy to promote the smart grid in the UK by indicating areas where DSR or energy storage would be most beneficial for the energy system (Elam 2016). Policy could then more accurately target the barriers and incentives required to enable the energy industry to develop a smart and flexible network (DECC 2015c). There is also scope for data to be used by a range of academic research seeking to improve energy efficiency and carbon reduction programmes and by initiatives that link energy and health outcomes, such as projects seeking to ensure vulnerable households do not under-heat their homes.

Notwithstanding these valuable national public interest uses of smart meter data, there is likely to be a range of uses of data that either will need to be enacted at a local scale or will benefit a specific section of society such as a geographic community. In particular, energy efficiency and fuel poverty programmes are, in practice, delivered locally and local partnerships tend to be central to success. Many local authorities, NGOs, and health and wellbeing organisations have a long history of working together to deliver local energy programmes, often through Affordable Warmth Partnerships. At this scale, the spatial and temporal resolution of smart meter data could vastly improve the tailoring of energy advice, the identification of the fuel poor, and the matching and evaluation of measures and it is difficult to envisage effective use of smart metering data to improve energy efficiency and fuel poverty that does not engage local actors in delivery.

Social landlords are often already involved in local energy efficiency and fuel poverty programmes and smart meter data could offer new opportunities for ensuring that vulnerable tenants do not live in under-heated properties. Similarly, health providers are already involved in fuel poverty referral schemes and smart meter data could be integrated with other health data.

Many of these uses of smart meter data would take place at a city scale and resonate with the "smart cities" agenda more generally. As discussed, both the term "smart city" and role of smart cities in addressing modern social challenges are contested; however, cities are increasingly seeking to incorporate energy into wider "smart city" agendas and strategically plan local energy systems. These approaches generally bring together many of the smart energy data uses already discussed in to provide local, public interest focussed, coordination of smart energy data.

Indeed, a number of cities are constructing smart city visions which frame city-scale approaches to smart energy data as a route to maximising the public interest benefits of data and are explicitly exploring if "the city" is a viable organising scale for the management and application of smart energy data. For example, the "Bristol Smart Energy City Collaboration" is aiming for Bristol to be the UK's first smart energy city and has a vision that by 2020, "Bristol will have a public-interest organisation coordinating the smart use, distribution and supply of heat and power across the city for the benefit of its people and businesses" (Centre for Sustainable Energy 2015, p. 6). The partnership includes a range of organisations such as the Centre for Sustainable Energy, Bristol City Council, Bristol University, Western Power Distribution, Arup and KPMG, and seeks to explore how the benefits of smart energy data can be organised at a city scale.

In addition to this explicit "city" focus, there is also increasing interest in how community scale public interest organisations may be able to utilise smart meter data, including community energy groups using data to target, monitor and evaluate their programmes. For example in Wadebridge, Cornwall a community energy group (Wadebridge Renewable Energy Network (WREN)) worked with Western Power Distribution, Tempus Energy and Regen SW to explore the potential for an offset connection agreement to shift local consumption to when photovoltaics (PV) are generating. The scheme links community-owned PV generation with a local "sunshine" tariff to reduce grid congestion (WREN 2015) with smart meter data central to the model.

There is also increasing interest from a number of local authorities in taking a more central role in the energy system through the establishment of municipally owned energy supply companies (Platt *et al.* 2014, Ofgem 2017). These models tend to actively cite a desire to ensure that the (local) public interest is served through either operating on a not-for-profit basis or reinvesting profits into sustainable energy projects. Examples include Robin Hood Energy (Robin Hood Energy 2016) – a not-for-profit energy company wholly owned by Nottingham City Council; Bristol Energy (Bristol Energy 2016) – a municipally owned energy supply company; and Our Power (Our Power 2016) – a not-for-profit supplier founded by a consortium of Scottish housing associations and local authorities. As well as developing full energy supply undertakings a number of local authorities are exploring partnership approaches to energy supply such as through "OVO Communities".[7] Whilst these city-scale energy supply models do not necessarily rely on access to smart meter data, most have ambitions to access smart meter data to provide additional services to customers and to access new sources of revenue. For example, Robin Hood Energy is offering new customers smart meters and is exploring using smart meters to deliver better tariffs for prepayment customers.

Overall, it is evident that a number of sub-national public interest uses of smart meter data may be possible within the current arrangements. There is also increased interest from cities in engaging in the energy system through (not-for-profit) energy supply models. Moreover, there is potential for smart meter data to support the development of the smart city agenda, improve the targeting and monitoring of local energy programmes, support social housing and health services, provide new opportunities for community energy and support other local approaches to balancing supply and demand.

3.4. Challenges for city-scale public interest uses

As the previous section illustrates, there is a range of potential public interest uses of smart meter data and it is likely that there will be a number of public interest uses where local actors play an important role. Despite this, there is little mention of public interest benefits in smart metering discourses and even less debate of whether cities and other local actors could play an important role in realising public interest benefits. This section outlines the issues which might impact on public interest uses of smart meter data (particularly at the city scale) under the current smart metering framework and focusses on issues of complexity, incumbency, data access and aggregation.

3.4.1. Complexity and incumbent advantage

The GB smart metering DAPF rightly has consumer choice at its heart. However, the resulting privacy and access requirements (as outlined in Section 3.2) are highly complex and are likely to involve significant costs for small organisations wishing to engage in smart meter data. This may have a detrimental impact on public interest uses of smart meter data as many public interest users of smart meter data, and certainly most city-scale public interest uses, are likely to have limited resources to engage in this complexity or meet these costs.

The current framework may also favour incumbent energy suppliers as these organisations are managing the roll out of smart meters and already have systems in place to manage ongoing interactions with customers. Non-incumbent, particularly locally based, interests may therefore have little opportunity to engage in the smart metering agenda due to resourcing and knowledge deficit issues. There is also limited acknowledgement of the potential for city-scale public interest uses of smart metering data by existing smart metering institutions and actors. So despite much activity by city-scale public interest actors to explore how they can engage in smart meter data it appears that the agenda is currently dominated by incumbent interests (such as utilities, network operators, data specialists and Ofgem) with limited space for the local state, civil society or other socially and environmentally focussed groups to engage.

3.4.2. Data access

One of the main barriers for public interest uses of smart meter data relates to data access. For example, DCC users need to follow the same application process regardless of the uses of data they wish to make, so a small community data user would have to follow the same processes as a large national supplier.

Additionally, within the existing DAPF the "third party" data user category is very broadly defined, including any non-supplier or non-DNO use of smart data – ranging from large data management companies to local community groups. Reference to community groups, in particular, was omitted from the DAPF despite the differing ability of community groups and other small/not-for-profit organisations to engage in complex data management processes. Specifically, the strict consent requirements and rigorous auditing processes of the DCC and SEC are likely to create barriers for these city-scale actors (Marshall 2015).

More broadly, current third-party data access options are likely to raise barriers which are particularly pertinent to city-scale public interest actors. As discussed, there are currently three options for third parties to access smart meter data; (1) as a registered DCC User and SEC signatory; (2) via a commercial contract with an existing DCC user; or (3) directly from a CAD with the customer's permission. Although there is some uncertainty regarding the data access options public interest users of smart meter data might choose, the cost of providing and installing CADs is likely to be relatively high so it seems likely that most organisations who wish to access smart meter data will do so via the DCC Gateway (directly or via a third party). If organisations choose to access data directly, then they will be subject to significant costs and complexity relating to becoming a DCC user and a SEC signatory, which are likely to be particularly significant for city actors such as local authorities and community groups. If organisations wish to access data via a contract with an existing DCC user, then there is currently no precedent for how these arrangements might operate and a lack of clarity as to whether existing DCC users might share these data at an acceptable cost.

3.4.3. Aggregation

There is also a lack of debate in current smart meter policy discourses regarding how organisations might access half-hourly smart meter data on a locally aggregated basis. This might be particularly important for a number of city-scale public interest uses of smart meter data, such as local energy efficiency programmes, which may only need aggregated data for a block of flats or social housing estate as the physical characteristics of individual properties are similar.

Additionally, half-hourly data aggregated at the substation or low voltage feeder level would be sufficient to indicate areas of grid constraint for many DSR uses of smart meter data. These data would potentially be useful to DNOs, local authorities, social housing providers and community energy projects. Currently, DNOs hold data on total demand and maximum demand at this scale but these data are not available at a half-hourly resolution, limiting its use for demand management purposes. Sourcing this locally aggregated data direct from suppliers would not be viable as it is unlikely that one supplier will hold comprehensive local data (as they would not be the supplier for 100% of residents). It is also probably that cost and practicality issues would limit the ability of local public interest organisations to access these data directly from consumers.

4. Discussion and proposed research agenda

4.1. Potential for city-scale public interest benefits

This review suggests that there is significant potential for smart meter data to benefit the public interest but these benefits are under-represented in smart meter discourses. The current smart meter data framework does little to recognise the potential collective, as well as individual, benefits of smart meter data or the unique opportunities and barriers facing city-scale public interest uses.

In drawing these conclusions, it is important to note that the benefits and barriers to public interest uses of smart meter data are, as yet, largely unquantified and activity by public interest organisations in this space tends to be characterised by a high level of experimentation and aspiration. Notwithstanding this uncertainty, the public interest benefits of city-scale uses of smart meter data are potentially significant but smart metering discourses have been dominated by individual consumer gains and the benefits new commercial uses of smart meter data might bring. Whilst these uses are important, they do not account for the fact that some people will be more able and more likely to utilise their smart meter data and there is, therefore, potential for smart metering benefits to be unevenly distributed across society. This has important implications in relation to the equity of smart energy systems as the cost of the smart meter roll out is being borne by all energy consumers.

For the full range of public interest benefits arising from smart metering data to be realised market rules and regulations would need to develop which do not explicitly disadvantage small, local or community based models. This may include ensuring that partnership-based supply models are accommodated in smart meter data access and privacy arrangements or that DCC and SEC arrangements are not prohibitively complex or costly for sub-national public interest users. Addressing these barriers to public interest uses of smart meter data, and acknowledging that many of these uses will be realised at city or community scale, may contribute to the equity of the future smart energy system through the promotion of an inclusive energy system.

On a practical level, a range of measures could begin to address the barriers to achieving public interest benefits. For example, Government and regulatory agencies could set out how they will engage with public interest users of smart meter data in their ongoing work, particularly at the sub-national level. This could take the form of an advisory group on smart meter data and the public interest or support to increase the participation of public interest bodies on SEC panels, committees and working groups.

Non-incumbent actors could also be supported through greater provision of collective resources for public interest users of smart energy data. Further work would be required to clarify what resources might be most appropriate but they could include a smart meter data hub which acts as a data consent repository and tracks energy consumer permissions regarding the use of their smart meter data. Such a hub could also act as an interface with the DCC Gateway to manage third-party public interest access to smart meter data and provide additional management, processing and preparation of data in order to provide it to public interest users in useful formats.

Access to smart meter data by a wider range of public interest organisations could also be facilitated if some data were made open access. The open publication of data is increasingly being framed as an important element of smart cities (Ojo *et al.* 2015) and there is emerging interest in energy open data, such as the Amsterdam Energy Atlas (Amsterdam Smart City 2016) which provides open data on solar potential, waste heat and other urban energy data sources, and Open Data Bristol which provides a range of energy-related open data on fuel poverty, solar potential and the energy consumption of city council buildings. Clearly, there are consumer privacy implications to making any domestic smart meter data open access; however, an exploration of the extent to which arrangements for anonymization and aggregation can facilitate this may be useful.

4.2. The intersection of urban governance literature and transitions approaches

The analysis of the GB smart meter data framework in this paper suggests that although there are likely to be a range of opportunities for different actors to utilise smart meter data, city-scale public interest uses may experience some unique challenges with the current framework privileging national and commercial approaches over the sub-national.

A great deal of urban governance literature focusses on the potential for smart city agendas to be dominated by neoliberal forms of urban development; however, this paper suggests that there is equally potential for smart city approaches to smart meter data to enable the realisation of increased public interest benefits. The analysis also indicates that the ability of discourses focussed on public interest benefits and social equity to co-exist with business-led discourses may be limited by the national regime for smart meter data. This highlights that although there may be the potential for cities to play a role in realising a just and socially equitable energy transition they may be limited in their ability to shape (local) technical transitions by the configuration of national regimes.

These conclusions acknowledge that urban low-carbon socio-technical experiments often blur the distinction between public and private authority creating "new forms of political space … enacted through forms of technical intervention in infrastructure networks" (Bulkeley and Castán Broto 2013, p. 361). In this context, there is potential for city-scale approaches to smart meter data to be dominated by commercial interests and navigating uses of smart energy data to achieve public interest benefits is far from unproblematic. There is, therefore, also a need to recognise that many cities are adopting an entrepreneurial approach to smart cities and explore in greater detail the position and influence of the private sector in shaping local smart city agendas.

It is also evident that cities might be constrained in their ability to challenge the dominance of incumbent actors in smart meter discourses due to the outsourcing of IT services and a lack of data management skills leading to limited ability for local government to intervene and a reduced "quality of knowledge and knowing within the state" (Taylor Buck and While 2015, p. 502). This might be particularly difficult as, in addition to the dominance of incumbent interests in the national smart meter framework, there may also be interests operating at the city scale which promote the adoption of market-led solutions to urban governance (Kitchin 2015).

At the city scale, there is often much debate of the potentially transformative power of smart city concepts; however, there is much less focus on how smart cities will be enacted, by whom and how the public interest will be protected and enhanced (Hodson and Marvin 2014). Other authors have discussed the potential for the promised benefits of smart cities not to be realised (Luque *et al.* 2014) and the risk that citizens are marginalised in the construction of smart cities with private and commercial interests dominating the agenda (Hollands 2008). This paper demonstrates that a number of cities are interested in exploring the potential of smart meter data to support "smart city" agendas and to deliver collective, public interest benefits but there is a need for greater exploration of the competing interests operating at this scale and of how cities are operationalising smart energy data.

Both Hodson and Marvin (2012) and Dewald and Truffer (2012) emphasise that city actors can play a significant role in shaping processes of socio-technical change but that innovations at this scale are also intricately intertwined with, and potentially constrained by, actions at other (regional, national and international) scales. This review supports these assertions, identifying the city as an important arena for the delivery of public interest uses of smart meter data and highlighting how national regimes can shape capacity to act at this level. To explore these issues more fully, there is a need for further interrogation of the role of cities as sites of low-carbon experimentation in relation to smart energy data (Bulkeley and Castán Broto 2013, Carvalho 2015) and of how novel actor networks are (or are not) forming, including the role of city-scale intermediaries (Hodson *et al.* 2013). It is also important to note both the role of national regimes in limiting sub-national smart meter experimentation and the potential for city approaches to be dominated by marketised and entrepreneurial priorities. Both these issues are important areas for further research on the intersection of urban transitions and the equity of smart energy systems.

At a more detailed level, as Carvalho *et al.* (2012) identify, the ability of cities to engage in socio-technical transition processes partly depends on local competencies and institutional structures. Given that cities are likely to be limited in their ability to engage in smart energy activities, future research could usefully explore under what circumstances local approaches to public interest uses of smart meter data are being pursued. As an increasing number of smart city approaches engage in smart energy data what influences the distribution of frontrunners and laggards? What are the (not least equity) implications if public interest uses of smart meter data develop in a geographically uneven manner? What issues if more equitable uses of smart meter data develop, at least partly, through processes of experimentation?

Finally, this review also demonstrates the socio-technical nature of smart metering developments with institutional and governance structures of equal importance to the deployment of the technical measures (such as smart meters, smart grids and data management systems). Whilst the need for technological changes to co-evolve with institutions, practices, business models, actor networks and cultural norms (such as privacy expectations) (Carvalho 2015) is not controversial, this case demonstrates the dominance of technological perspectives in the GB smart metering governance regime, with physical, functional, technical and privacy aspects emphasised over social equity concerns.

5. Conclusion

This review suggests that there is significant potential for smart meter data to benefit the public interest, but these benefits are under-represented in smart meter discourses in GB. In particular, the unique opportunities and barriers facing city-scale public interest uses are not acknowledged. City-scale actors may face difficulties in seeking to use smart meter data in the public interest due to the dominance of incumbent, large-scale actors in shaping the agenda to date. The difficulties for decentralised solutions operating within the existing locked-in centralised energy system have been well documented (Unruh 2000, Meadowcroft 2009), but this work suggests that there is a high risk that such challenges remain as the smart energy system develops.

The availability of access to smart meter data is also taking place at the same time as a surge of interest in the involvement of city-scale actors in the energy system, particularly local authorities. As Fudge *et al.* (2016, p. 4) suggest "the role and influence of local authorities in the sustainable energy transition has often been either underplayed or they have been viewed as a part of the 'dominant regime'"; however, in relation to smart meter data is seems that the issues and opportunities faced by local authorities have more in common with community energy uses and do not constitute part of the dominant regime. There is, however, the potential for city-scale approaches to smart meter data to be operationalised in a number of ways with competing collective, individual, public and private interests driving differing outcomes. This paper highlights that numerous smart meter data public interest benefits could be enacted at a city scale, but the current framework privileges national and commercial approaches over sub-national. There is a need to further explore competing interests at this scale and a number of other areas for research are proposed.

Additionally, this paper indicates that the ability of alternative discourses – focussed on public interest benefits and social equity – to co-exist with business-led discourses may be limited by the national regime for smart meter data. This highlights that although there may be the potential for cities to play a role in realising a just and socially equitable energy transition they may be limited in their ability to shape (local) technical transitions by the configuration of national regimes.

Finally, more broadly, the findings of this review resonate with the research agenda outlined by van der Horst *et al.* (2014) in their introduction to a special issue on smart metering technology and society. In particular, they highlight the need to address questions of the configuration of control, ownership and management of smart technologies and the negotiation and governance of benefit sharing. Similarly, this paper emphasises that the distribution of benefits and control in relation to smart meter data may vary greatly based on different models of delivery and suggests that far more critical analysis is required of the interplay between multiple scales of governance in the social and political shaping of smart energy systems.

Notes

1. For example, EU Member States are required to roll-out electricity smart meters to 80% of consumers by 2020 (where long-term cost-benefit analysis is positive) (European Commission 2014) and over 57 million residential smart meters have been installed across the United States (US Energy Information Administration 2017).
2. TEDDINET – Transforming Energy Demand through Digital Innovation NETwork – is an academic research network addressing the challenges of transforming energy demand, as a key component of the transition to an affordable, low carbon energy system. Funded by the UK EPSRC (Engineering and Physical Sciences Research Council), TEDDINET's primary purpose is to share knowledge and enhance the impact of existing research.
3. As both the loci for radical innovations and the established rules and practices that stabilize existing socio-technical systems (termed niches and regimes, respectively, in the language of the multi-level perspective) (Geels 2011).
4. There is some debate as to whether it is likely that the role out will be completed by the end of 2020 (House of Commons: Energy and Climate Change Committee 2015).
5. Such as through time of use tariffs.
6. A Consumer Access Device (CAD) is a device that connects to the smart meter and provides a range of functionality, including collecting and distributing electricity and gas data. Additional functionality could involve the collection of additional (non-meter) data such as temperature and humidity, and interaction with smart appliances and smart boilers.
7. "OVO Communities" aims to develop partnerships between local authorities, housing associations, community groups and OVO to supply energy through white label arrangements. It aims to connect up local power generation, energy supply, energy efficiency and smart metering (OVO 2015).

Acknowledgements

This paper is indebted to the TEDDINET research network, Sustainability First and the Centre for Sustainable Energy, who commissioned the research which informed this paper and gave their time to debate the issues. The paper also benefited from insightful comments from two anonymous reviewers.

Disclosure statement

No potential conflict of interest was reported by the author.

Funding

This work was supported by a TEDDINET research challenge on smart meter data and the public interest. Funded by the UK EPSRC (grant number EP/L013681/1) TEDDINET is an interdisciplinary research network addressing the challenges of transforming building energy demand as part of the low-carbon transition.

ORCID

Jess Britton http://orcid.org/0000-0001-5302-0804

References

Amsterdam Smart City, 2016. Energy atlas [online]. Available from: https://amsterdamsmartcity.com/projects/energy-atlas [Accessed 17 July 2017].
Bristol Energy, 2016. Welcome to Bristol Energy [online]. Available from: https://bristol-energy.co.uk/ [Accessed 10 April 2017].
Britton, J., 2016. *Smart meter data and public interest issues – the sub-national perspective*. Bristol: Centre for Sustainable Energy.
Bulkeley, H. and Castán Broto, V., 2013. Government by experiment? Global cities and the governing of climate change. *Transactions of the Institute of British Geographers*, 38 (3), 361–375.
Carvalho, L., 2015. Smart cities from scratch? A socio-technical perspective. *Cambridge Journal of Regions, Economy and Society*, 8 (1), 43–60.
Carvalho, S., Mingardo, G., and Haaren, J.V.A.N., 2012. Green urban transport policies and cleantech innovations: evidence from Curitiba, Göteborg and Hamburg. *European Planning Studies*, 20 (3), 375–396.
Centre for Cities, 2014. *Smart cities*. London: Centre for Cities.

Centre for Sustainable Energy, 2015. *Bristol smart energy city collaboration. Towards a smart energy city: mapping a path for Bristol. Reflections from 2015.* Bristol: Centre for Sustainable Energy.

Coenen, L. and Truffer, B., 2012. Places and spaces of sustainability transitions: geographical contributions to an emerging research and policy field. *European Planning Studies,* 20 (3), 367–374.

Darby, S., 2010. Smart metering: what potential for householder engagement? *Building Research & Information,* 38 (5), 442–457.

DECC, 2012a. *Smart metering implementation programme: data privacy and security. Government response to consultation.* London: Crown Copyright.

DECC, 2012b. *Smart metering implementation programme: privacy impact assessment.* London: Crown Copyright.

DECC, 2014a. *DECC smart meter impact assessment final January 2014.* London: Crown Copyright.

DECC, 2014b. *Smart metering implementation programme. Smart metering equipment technical specifications.* London: Crown Copyright.

DECC, 2015a. *DECC smart metering implementation programme. Government response to consultation on timing of the review of the data access and privacy framework.* London: Crown Copyright.

DECC, 2015b. *Smart meters, smart data, smart growth.* London: Crown Copyright.

DECC, 2015c. *Towards a smart energy system.* London: Crown Copyright.

Dewald, U. and Truffer, B., 2012. The local sources of market formation: explaining regional growth differentials in German photovoltaic markets. *European Planning Studies,* 20 (3), 397–420.

Elam, S., 2016. *Smart meter data and public interest issues. Discussion paper 1: national perspective.* Bristol: Centre for Sustainable Energy.

European Commission, 2014. *Cost-benefit analyses & state of play of smart metering deployment in the EU-27.* Brussels, Belgium: European Commission.

Fudge, S., Peters, M., and Woodman, B., 2016. Local authorities as niche actors: the case of energy governance in the UK. *Environmental Innovation and Societal Transitions,* 18, 1–17.

Geels, F.W., 2011. The multi-level perspective on sustainability transitions: responses to seven criticisms. *Environmental Innovation and Societal Transitions,* 1 (1), 24–40.

Glasmeier, A. and Christopherson, S., 2015. Thinking about smart cities. *Cambridge Journal of Regions, Economy and Society,* 8 (1), 3–12.

Goodspeed, R., 2015. Smart cities: moving beyond urban cybernetics to tackle wicked problems. *Cambridge Journal of Regions, Economy and Society,* 8 (1), 79–92.

Hills, J., 2012. *Getting the measure of fuel poverty: final report of the fuel poverty review.* London: Crown Copyright.

Hodson, M. and Marvin, S., 2009. Cities mediating technological transitions: understanding visions, intermediation and consequences. *Technology Analysis & Strategic Management,* 21 (4), 515–534.

Hodson, M. and Marvin, S., 2010. Can cities shape socio-technical transitions and how would we know if they were? *Research Policy,* 39 (4), 477–485.

Hodson, M. and Marvin, S., 2012. Mediating low-carbon urban transitions? Forms of organization, knowledge and action. *European Planning Studies,* 20 (3), 421–439.

Hodson, M. and Marvin, S., 2014. Introduction. *In*: M. Hodson and S. Marvin, eds. *After sustainable cities.* Abingdon: Routledge, 1–9.

Hodson, M., Marvin, S., and Bulkeley, H., 2013. The Intermediary Organisation of low carbon cities: a comparative analysis of transitions in Greater London and Greater Manchester. *Urban Studies,* 50 (7), 1403–1422.

Hollands, R.G., 2008. Will the real smart city please stand up? *City,* 12 (3), 303–320.

van der Horst, D., Staddon, S., and Webb, J., 2014. Smart energy, and society? *Technology Analysis & Strategic Management,* 26 (10), 1111–1117.

House of Commons: Energy and Climate Change Committee, 2015. *Smart meters: progress or delay? Ninth report of session 2014–15.* London: Crown Copyright.

De Jong, M. *et al.,* 2015. Sustainable-smart-resilient-low carbon-eco-knowledge cities; making sense of a multitude of concepts promoting sustainable urbanization. *Journal of Cleaner Production,* 109, 25–38.

Kitchin, R., 2015. Making sense of smart cities: addressing present shortcomings. *Cambridge Journal of Regions, Economy and Society,* 8 (1), 131–136.

Lawhon, M. and Murphy, J.T., 2012. Socio-technical regimes and sustainability transitions: insights from political ecology. *Progress in Human Geography,* 36 (3), 354–378.

Luque, A., McFarlane, C., and Marvin, S., 2014. Smart urbanism: cities, grids and alternatives? *In*: M. Hodson and S. Marvin, eds. *After sustainable cities?* Abingdon: Routledge, 74–90.

Marshall, E., 2015. *Use of smart metering data to enhance the delivery of energy efficiency policies.* Working papers of the Sustainable Society Network, Vol. 5. London: Imperial College London.

Meadowcroft, J., 2009. What about the politics? Sustainable development, transition management, and long term energy transitions. *Policy Sciences,* 42 (4), 323–340.

Ofgem, 2017. *Local energy in a transforming energy system. Ofgem's future insights series.* London: Ofgem.

Ojo, A., Curry, E., and Zeleti, F.A., 2015. A tale of open data innovations in five smart cities. *In*: Proceedings of the 2015 48th Hawaii International Conference on System Sciences (HICSS). Grand Hyatt, HI, 5-8 January 2015. New York: IEEE.

Our Power, 2016. Our Power [online]. Available from: http://our-power.co.uk/ [Accessed 24 April 2017].

OVO, 2015. *Say hello to OVO communities*. Bristol: OVO.

Platt, R. *et al.*, 2014. *City energy: a new powerhouse for Britain*. London: IPPR.

Robin Hood Energy, 2016. Robin Hood Energy [online]. Available from: https://www.robinhoodenergy.co.uk/ [Accessed 24 April 2017].

Rutgers, M.R., 2015. As good as it gets? On the meaning of public value in the study of policy and management. *The American Review of Public Administration*, 45 (1), 29–45.

Rutherford, J. and Coutard, O., 2014. Urban energy transitions: places, processes and politics of socio-technical change. *Urban Studies*, 51 (7), 1353–1377.

Smith, A., Voß, J.-P., and Grin, J., 2010. Innovation studies and sustainability transitions: the allure of the multi-level perspective and its challenges. *Research Policy*, 39 (4), 435–448.

Sustainability First, 2015. *Towards a definition of the long-term public interest. A new-pin background working paper*. London: Sustainability First.

Taylor Buck, N. and While, A., 2015. Competitive urbanism and the limits to smart city innovation: the UK future cities initiative. *Urban Studies*, 54 (2), 1–33.

Unruh, G.C., 2000. Understanding carbon lock-in. *Energy Policy*, 28, 817–830.

US Energy Information Administration, 2017. How many smart meters are installed in the United States, and who has them? [online]. Available from: https://www.eia.gov/tools/faqs/faq.php?id=108&t=3 [Accessed 25 April 2017].

Viitanen, J. and Kingston, R., 2014. Smart cities and green growth: outsourcing democratic and environmental resilience to the global technology sector. *Environment and Planning A*, 46(4), 803–819.

WREN, 2015. Sunshine Tariff Trial [online]. Available from: http://www.wren.uk.com/sunshine [Accessed 25 April 2017].

Stretching "smart": advancing health and well-being through the smart city agenda

Gregory Trencher ⓘ and Andrew Karvonen ⓘ

ABSTRACT

Contemporary smart cities have largely mirrored the sustainable development agenda by embracing an ecological modernisation approach to urban development. There is a strong focus on stimulating economic activity and environmental protection with little emphasis on social equity and the human experience. The health and well-being agenda has potential to shift the focus of smart cities to centre on social aims. Through the systematic and widespread application of technologies such as wearable health monitors, the creation of open data platforms for health parameters, and the development of virtual communication between patients and health professionals, the smart city can serve as a means to improve the lives of urban residents. In this article, we present a case study of smart health in Kashiwanoha Smart City in Japan. We explore how the pursuit of greater health and well-being has stretched smart city activities beyond technological innovation to directly impact resident lifestyles and become more socially relevant. Smart health strategies examined include a combination of experiments in monitoring and visualisation, education through information provision, and enticement for behavioural change. Findings suggest that smart cities have great potential to be designed and executed to tackle social problems and realise more sustainable, equitable and liveable cities.

1. Introduction

Visions and efforts to realise sustainable cities over the past three decades have largely focused on the simultaneous pursuit of economic development and environmental protection (Hodson and Marvin 2017). The current smart city agenda has embraced this ecological modernisation approach to urban development. By weaving "smart" information communication technologies (ICT) into the urban landscape, smart cities promise to improve environmental and economic performance through data collection, analysis and evidence-based policy-making (Goodspeed 2015, March in press). This can enable more intelligent planning and optimised or novel service delivery by munici-palities and industry (Saujot and Erard 2015). In parallel, real-time data from smart devices can guide citizen behaviours and choices regarding matters such as home energy use. Meanwhile, empirical studies of smart city projects reveal that the vast majority of projects are overwhelmingly fixated on driving physical environment and infrastructure improvements and fostering innovation in service of economic growth (Albino *et al.* 2015, Alizadeh 2017). Largely ignored, on the other hand, is the human experience and the possibility of contributing to improved well-being and urban livelihoods through application of smart technologies.

The process of urban development represents an important opportunity to pursue the social dimensions of city life such as health and well-being. Health is a long-standing urban agenda. Nowhere is this exemplified better than the Sanitary City movement of the late nineteenth and early twentieth centuries (Melosi 2000, Pincetl 2010, Karvonen 2011). Faced with countless deaths from transmissible diseases such as cholera and tuberculosis, city planners and engineers significantly improved human health in modern cities by introducing systems of water supply, wastewater treatment and solid waste collection. In recent times, the link between health and the urban environment has taken on a renewed light through emerging scholarship. Söderström (2016) has documented interesting cause-and-effect linkages between dense and chaotic urban environments and mental health while Thomas *et al.* (2014) argue that well-being benefits can flow from positive mental health in cities.

Propelled by breakthroughs in technology and data science, the agenda of health and well-being is increasingly becoming an attractive target for digital innovation. The development of digital health records, the introduction of remote doctor visits and the rise of wearable consumer devices (e.g. FitBit, Apple Watch) herald an emerging digital age for medical services. Moreover, health care is a costly but essential public service. Many governments around the world are therefore keen to leverage ICT to improve the health of citizens and boost the effectiveness of health-related services to reduce the burden of health care expenditures on the public purse (Goodspeed 2015). Cognizant of such trends, scholars have recently highlighted a yet to be fulfilled potential for the smart city agenda to contribute to the enhancement of health and well-being of citizens (Haarstad 2016, March and Ribera-Fumaz 2016, Meijer and Bolívar 2016, Bibri and Krogstie 2017). However, as highlighted by Haarstad (2016), visibly lacking is empirical scholarship that actually demonstrates this potential. A stretched smart city agenda exploiting digital technologies to improve resident health and well-being represents a stark contrast to the techno-utopian agendas of Masdar City, Songdo and other exemplar smart cities around the world. Scholars have critiqued such projects for their narrow pursuit of technological innovation in service of economic growth and selected corporate interests and their failure to tie the trial and diffusion of technologies to messy social issues and concrete goals of improving resident livelihoods (Hollands, 2015). The potential for an expanded conception and execution of the smart city thereby points to the possibility of using technologies not as ends in themselves – as is most often the case (Glasmeier and Nebiolo 2016) – but rather, as a means to improve the social aspects of urban living.

While scholarship on the adoption of ICT to pursue greater health and well-being through smart city development is nascent, Japan is rapidly emerging as a global frontrunner in this field. Prompted by an accelerating transition to an aging population, private, public and third sector actors in a handful of urban development projects are ambitiously stretching dominating conceptions of smart cities by actively experimenting with ICT and smart planning to advance preventative health and longevity. Addressing this trend, this study examines on-the-ground experiences and accumulated learning in Kashiwanoha, a national (and possibly global) pioneer of smart city health strategies. Through a case study approach, the primary objective of this article is to increase understanding around the potential for smart cities to move beyond conventional environmental and economic agendas to tackle a social challenge such as the pursuit of healthier lifestyles and greater well-being. This paper contributes to the emerging social agenda in smart city discourse (Glasmeier and Christopherson 2015, Goodspeed 2015, Stollmann *et al.* 2015, Glasmeier and Nebiolo 2016, Bibri and Krogstie 2017) and addresses the following research questions:

- How is health envisioned in the smart city and how is it connected to other sustainability agendas?
- What approaches are used to advance health and well-being of residents and what outcomes, challenges and learning have resulted?

The following background section provides an overview of the current smart city agenda and its failure to address social issues. We then introduce digital health and well-being as an approach to address this gap by focusing on social welfare, resident well-being and lifestyles. Our case study of Kashiwanoha then demonstrates how in practice a smart city can envision and implement an array of technology- and people-centred measures to guide residents towards greater health and well-being. Findings demonstrate an array of top-down and bottom-up approaches including experiments in monitoring and visualisation, public education through digital and face-to-face information provision and the creation of intrinsic and extrinsic incentives from ICT devices to spur behavioural change. These approaches represent a wide range of urban stakeholder configurations and differing degrees of technological and data dependency. Key outcomes of these initiatives include enhanced community building from joint learning, and resident participation around shared values of preventative health. Digital tools are also providing novel opportunities for municipal actors and health professionals to communicate more effectively with residents around health issues. The experiences in Kashiwanoha also highlight potential obstacles to a smart health agenda that revolve around data privacy issues, commercialisation opportunities and engaging with a large sector of the population. While Kashiwanoha's smart health strategies have produced mixed levels of success, they provide novel insights into how smart city agendas can be stretched to encompass attempts to improve the social and well-being dimensions of urban life. Moreover, they demonstrate how technological innovation can be framed not as an end in itself but as a means to address localised social problems and enhance the lives of urban residents.

2. Background

2.1. Expectations, tensions and shortfalls of the smart city agenda

The majority of smart-sustainable projects focus on entrepreneurial forms of urban development that utilise cutting-edge technologies to simultaneously boost the economy while reducing environmental impacts (Viitanen and Kingston 2014, Glasmeier and Christopherson 2015, Hollands 2015, Glasmeier and Nebiolo 2016). This is the classic formula of green growth that has dominated sustainable urban development discourse for the last three decades (Schuurman et al. 2012, Tranos and Gertner 2012, Lee et al. 2014). As Haarstad (2016, p. 7) notes, "sustainability is largely an assumed result of more efficient, cost-effective urban systems and greater availability of data." Meanwhile, social considerations are limited to supporting job creation and encouraging citizen participation through open data platforms (Bakıcı, Almirall and Wareham 2013, Arrizabalaga et al. 2016, Lee et al. 2014, Hielkema and Hongisto 2013). Glasmeier and Nebiolo (2016, p. 2) contend that "smart" labels that we attach to everything from wireless sensors to automated vehicles are frequently unconnected to social objectives, arguing "the unintended consequence of smart city 'making' is to privilege technologies without equivalency tests that make clear what the public values are and what the basic needs are that these values seek to promote." This suggests that the starting point for smart cities should be social issues rather than the narrower goal of technology diffusion (Hollands 2015). In other words, technology-centred agendas should be demand-driven and focus on residents' needs rather than being supply-side driven and principally concerned with economic growth objectives (Söderström 2016). Thus, there is a need for smart cities to be more relevant to a broader array of societal issues (Glasmeier and Nebiolo 2016).

The objective of enhancing public health has the potential to address the social shortcomings of smart cities. Enhanced human health is increasingly seen as a co-benefit of urban planning, particularly in the field of public health (Giles-Corti et al. 2016). This is, however, principally from an urban engineering perspective where compact cities are designed to provide sustainable public transport, opportunities for walking, cycling, exercise, reduced crime, safe and nutritious food, vegetation in public spaces and clean air (Ramaswami et al. 2016, Sallis et al. 2016). Greater well-being and health for residents is produced by a cleaner and more liveable urban environment. While these physical environment planning approaches hold much promise to promote healthy lifestyles and positively impact the lives of urban residents as a whole, there is a flipside. Scholars like Caprotti et al.

(2015) underscore that the privilege of residing in corporation-driven new smart cities can constitute a luxury out of reach from poorer populations due to high property premiums.

Developments in data management and health care suggest that digital technologies have significant promise to advance human health and well-being in urban settings. In conventional smart cities, the Internet of Things (IoT) and ICT sensors link appliances, building components, transport systems and residents to increase efficiency of energy usage and allocation of resources. Similarly, emerging research and experiments in medicine and public health demonstrate that digital technologies can also link residents and technological artefacts with data and information networks to optimise health care efficiency and effectiveness (Andreassen *et al.* 2015). For example, wireless sensors can measure physical activity and provide data-based diet and lifestyle guidance (Koch 2010), electronic communication networks can link health care professionals, patients and family caregivers to enhance health literacy and preventative care (Haluza and Jungwirth 2015), "telemedicine" can enable homecare of the elderly and relieve doctor shortages (Obi *et al.* 2013) while social media applications and digital devices can promote the social inclusion of care recipients (Hasan and Linger 2016). Moreover, Thomas *et al.* (2014) argue that dominant conceptions of urban health need to be expanded beyond illness and disease to encompass mental and social well-being. This points to an important challenge for smart cities – that of using technologies to promote not only physical health but also mental and social well-being.

Recently, interest is mounting around the potential to harness such ICT innovations to public health in smart city developments. As part of the U.K. Smart Cities Forum, the City of Leeds and the University of Leeds advocated a vision and set of national government recommendations for a "smart city for health and wellbeing". The authors propose that such a city

> … takes an integrated approach to the use of digital technologies to increase the connectedness of people to the information and city functions that improve health and wellbeing, reduce inequalities and support a higher quality of life for all its citizens. (Leeds City Council and Leeds 2015, p. 13)

This vision calls for a balanced focus on well-being, illness prevention and treatment. Digital technologies are framed as tools to empower residents to transgress the role of a passive patient and become "active participants" in a shared and digitalised health care system.

2.2. Smart cities and health in Japan

Japan is a world leader in smart city developments with an estimated 160 projects funded by the national government up to 2014 (Mah *et al.* 2013, Nyberg and Yarime 2017). The realisation of smart cities is a strategic priority of the national government and supported through the significant expenditure of public funds. DeWit (2013) notes that prior to the Fukushima disaster, trials of smart city technologies by Japanese corporations and officials were limited to a few pilot cities with the aim of developing an export market to position Japan as a global leader in the field. These projects were predicated on the continued dominance of nuclear power and centralised energy provision. However, the Fukushima nuclear disaster of 2011 and consequential shutdown of Japan's entire nuclear fleet shattered these assumptions. The national government rapidly refocused the smart city agenda on the domestic market and prioritised energy conservation and disaster resilience through the development of local renewable energy and smart grid projects (Nyberg and Yarime 2017, Yarime and Karlsson in press).

A few Japanese smart city initiatives focus on retrofitting existing cities while most involve the construction of entirely new districts or towns. In addition to a focus on boosting the disaster resilience of energy supplies through diverse and distributed renewable sources (Kono *et al.* 2016), the national smart city discourse explicitly states that ICT will improve the quality of life for urban residents by tackling social problems such as disaster resilience, population aging and stagnant economic conditions (Kijou and Rure 2014). Japanese smart city projects are frequently underwritten with generous government funding while their implementation is spearheaded by a handful of large corporations (Mah *et al.* 2013, Yarime and Karlsson in press) in partnership with municipalities who willingly open up public resources for private-sector exploitation. In the majority of cases, citizen involvement is largely absent (DeWit 2013, Kijou and Rure 2014) or limited to "a very specific and limited set of expression and actions" (Granier and Kudo 2016, p. 72). This situation echoes observations from scholars regarding global trends where the dominating corporate smart city model fails to translate technological innovation into the creation of novel participatory roles for citizens in urban governance (Hollands 2015).

However, the national issue of an ageing population in Japan has the potential to open up the smart city agenda to democratic engagement with urban residents. The country continues to struggle with the emerging economic consequences of the most rapidly greying population in the world due to a chronically declining birth rate and increased life expectancy (Muramatsu and Akiyama 2011, Shirahase 2015). Both governments and corporations are increasingly looking to ICT and smart innovation to reduce health care costs and mitigate a looming viscous cycle whereby, on one hand, health expenditures rise in accord with population greying trends while, on the other hand, a shrinkage of the young working population leads to a decrease in taxes for public services (Obi *et al.* 2013). Accordingly, a growing number of municipalities have started experimenting with ICT to promote public health. Since health ultimately concerns lifestyles and requires the buy-in and engagement of residents, this emerging development signals an interesting development in smart cities for strengthening their social agenda and participatory function. In other words, it is through health and well-being that the social dimensions of smart urban development have the potential to flourish in Japan.

3. Methods

The Japanese smart city of Kashiwanoha provides a real-world case study of how the health and well-being agenda is being addressed through smart urbanisation. Health and longevity has formed a central and explicit focus of the Kashiwanoha Smart City since its establishment in 2008. This long tenure of a smart health agenda thus provides an important opportunity to reflect on achievements to date. That said, the Kashiwanoha example is not intended as a model of best practice. Instead, it illustrates how smart urban development strategies and digital tools can be stretched or reinvented to potentially address localised social issues (in this case an aging society) alongside environmental and economic challenges.

The empirical findings are derived from qualitative data collected from primary and secondary sources. Primary data were collected between 2014 and 2017 through four site visits and semi-structured interviews with eight actors including the private developer, public health professionals from the municipality, university researchers, non-profit sector planners and volunteer health workers. Interviewees were selected to provide a range of perspectives and experiences on the smart health initiatives in Kashiwanoha. All interviews were conducted in Japanese by the first author in person (except one conducted via Skype). Interviews were recorded, transcribed and coded manually to identify major themes. Secondary data were derived from a desk-based study of newspaper and magazine articles, smart city promotional and explanatory materials, internal project documents and academic publications.

4. Findings

4.1. Background on Kashiwanoha smart city

Located in the City of Kashiwa, the Kashiwanoha Smart City was initiated with the opening of a new train line connecting Tokyo with the northern city of Tsukuba in 2005. The developer, Mitsui Fudosan, owned abundant undeveloped land in the station vicinity and this created an opportunity to develop an entirely new city from scratch. At present, Kashiwanoha is comprised of mixed-use developments on a 273-hectare plot that includes retail and dining facilities, a satellite campus for the University of Tokyo, a hotel, a hospital, commercial and venture incubator spaces and several high-rise residential buildings. Over the last decade, the smart city's population has grown from zero to about 6000 residents. By 2030, it is expected to reach around 30,000 residents. Property prices are relatively higher than nearby areas and thus, most residents are retired couples or wealthy middle-class and young families who commute to Tokyo for work. The promise of healthier living thus addresses itself principally to a wealthier segment of Japanese society although, as we outline later, many health-orientated initiatives offer access and benefits to a wider population in the surrounding area.

Although driven by private investment from the developer, Kashiwanoha Smart City has unfolded through intimate industry–university–government collaboration. The non-profit planning headquarters Urban Design Centre Kashiwa was established in 2006 to facilitate collaborative planning of the city and promote citizen participation (UDCK 2017). Spurred by this broader governance framework for cross-organisational collaboration and innovation, a culture of experimentation with technical and social approaches to societal problems has burgeoned in the new city (Trencher *et al*. 2015). In 2008, the developer teamed with the prefectural government and two local universities (the University of Tokyo and Chiba University) in a collaborative visioning process that lead to the proposal of an "International Campus Town". A 13 km^2 area encompassing Kashiwanoha station, two universities and government research institutes was designated as a "campus" or urban laboratory for research and implementation of academic knowledge. Three areas were designated as development priorities: environmental sustainability, high-tech business creation and health and well-being (Kurata *et al*. 2013).

Effectively serving as the common binding thread for pursuing these three objectives, smartness is interpreted principally from two dimensions. Firstly, by applying cutting-edge knowledge and innovation produced through university–government–industry collaboration to development of the city (Kurata *et al*. 2013), and secondly, by integrating ICT and digital technologies to various energy and lifestyle domains. In addition, the city has fixed a long-term goal of reducing CO_2 emissions in 2030 by 60% (compared to a Business as Usual scenario) by rolling out a smart grid, renewable energy generation and storage, energy efficiency and residential energy consumption visualisation. Kashiwanoha is aggressively marketed as a "smart city" and touted as a new Japanese model of urban development that is the fruit of cutting-edge innovation attained through industry–university–government collaboration (Mitsui Fudosan 2017).

4.2. The vision of a smart, healthy city

Beyond CO_2 emissions and energy efficiency, a core objective in Kashiwanoha is to enhance resident health and well-being. This goal attaches high expectations to the potential of ICT and mirrors the focus on smart technology and real-time visual feedback of building energy consumption to advance energy sustainability in the city. In addition, there is a distinct focus on preventative health; i.e. preventing illness by encouraging healthy lifestyle habits for diet, sleep and exercise, etc. The city's health and well-being narrative encourages residents to adopt a daily maintenance mind set and assume greater responsibility for their personal health. It touches on an array of lifestyles issues ranging from diet and sleep to walking and socialising. By doing so, the preventative agenda explicitly communicates an objective of reducing health-related expenditures for the local municipality by fostering healthy lifestyles that would eventually lead to lower reliance on post-illness consumption of medication and professional medical services. Although the aim is to address residents of all ages, Kashiwanoha's fixes an explicit goal of advancing longevity. It thus affords special consideration to promoting elderly health, neatly aligning with the national government's funding expectations that future-orientated urban development in Japan addresses the growing needs of a greying society. Addressing health and longevity today is therefore seen as an investment in social capital for tomorrow, when aging trends and associated burdens on public welfare will be further pronounced.

Beyond preventative health, there is a strong emphasis on fostering social cohesion. This is driven by growing realisations in the Japanese medical community that elderly citizens leading socially active lifestyles are more likely to avoid early deterioration of physical and mental capacities. In connection to this, one interviewed smart city planner emphasised how the pursuit of health and longevity is intended to encompass mental well-being, and secondly, to be "fun" and thereby transcending a narrow focus on physical health. From another perspective, core to the vision of Kashiwanoha's smart, healthy city is the developer's rhetoric that " … residents can become healthy just by living here", so explained other interviewees. This mirrors nicely the alluring effortlessness promoted by Kashiwanoha's marketing efforts, which similarly, promise that residents can lead an environmentally friendly lifestyle by simply residing in the city.

Several distinct motivations for the focus on health and longevity in Kashiwanoha emerged from interviews. The developer emphasised how the health identity serves to distinguish Kashiwanoha from other smart cities, potentially creating greater demand for real estate assets in a competitive marketplace and aging society. For the municipality, motivations are also financial since health expenses are rising annually due to an increasing number of elderly residents requiring medical care. Health promotion initiatives therefore represent an opportunity to mitigate this growing financial burden on the municipality and the shrinking workforce that contributes to the local tax-base. For university researchers, Kashiwanoha Smart City and the surrounding City of Kashiwa offer a myriad of opportunities to use the urban environment as a laboratory for emerging research agendas and to trial solutions for societal challenges in conjunction with willing residents and private and public actors.

4.3. Approaches to smart health and well-being

Empirical findings reveal three principal approaches to pursuing smart health and well-being in Kashiwanoha: experiments in monitoring and visualisation, education through information provision and enticement for behavioural change. Table 1 summarises the defining attributes of each approach, with the details of each unpacked in the following sections. Each is described individually

Table 1. Smart health approaches in Kashiwanoha.

Approach	Objectives	Target population	Key actors
Experiments in monitoring and visualisation	• Use visual feedback of health data to promote active lifestyles and health • Create new business opportunities for technology companies	• Motivated, active adults • Recruited residents and workers in Kashiwanoha	• Developer • Private industry • Municipality
Education through information provision	Transition to digital child health diary for: • Storing health data and enabling mutual learning across couples • Diffusing tailored health information from municipality and increasing accessibility	• Young parents and children • Recruited residents and workers in Kashiwanoha, then expanding to entire City of Kashiwa	• Municipality • Private industry
	Establishing walk-in preventative health care centre (Ashita) to educate public and engage residents	• Older residents • Open to all, regardless of residence	• Developer • Citizens • Private industry • University
Enticement for behavioural change	Incentivise daily walking by rewarding with purchase redeemable points (extrinsic incentive) in Sukoyaka Links initiative	• Motivated, active adults • All residents and workers in Kashiwanoha	• Developer • Private industry
	Raise exercise motivation/intensity through intrinsic incentives by: • Aggregating and disclosing data collected from individual activity recorders • Fostering a group mentality and sense of competition by ranking individual performance against cohort average and other competitors	• Motivated, active adults • Recruited residents and workers in Kashiwanoha	• Developer • Private industry

but it is important to note that there is significant overlap and synergies. Since 2012, this portfolio of approaches forms an interconnected and iterative strategy for fostering smart health, and continues to evolve.

4.3.1. Experiments in monitoring and visualisation

Several large-scale experiments were conducted in Kashiwanoha with wearable ICT devices. These generated and stored health and lifestyle data on a cloud server and then delivered this to participants as real-time, visual feedback through an Internet portal. These ICT experiments aimed to support preventative health by generating continuous data streams from an array of daily activities (sleeping, walking, running, working, stepping on weighing scales, etc.) and delivering results to participants as easy-to-understand graphs and metrics. One notable example of such experimentation was conducted over five weeks in 2013. In partnership with three medical-tech firms and the municipality, the developer recruited 150 residents to trial the following four types of technologies in differing combinations and then provide feedback to the developers:

1. Wearable ICT activity monitors to capture continuous lifestyle data (see Figure 1 top left);
2. Digital pedometer to record steps walked/ran;
3. Digital scales to record weight, body mass index (BMI) and body fat percentage (see Figure 1 bottom left), and;
4. A cloud server and Internet portal to store and display real-time data, and additionally, a forum to receive feedback and advice from municipality health professionals like nurses and dieticians.

Figure 1. Key ICT technologies and Ashita education centre. Top left: Wearable ICT activity recorder (image by permission from A&D Medical Co. Ltd. © 2017). Bottom left: ICT scales (image by permission from A&D Medical Co. Ltd. © 2017). Top right: Sukokaya Links programme pedometer and points card (image by author). Bottom right: Ashita walk-in health prevention educational centre (image by author).

Key visualisations derived from health data included graphs depicting shifts over the two-month period for steps walked/ran, exercise intensity (shown as Metabolic Equivalent), BMI and weight. As an attractive feature, participants in the groups wearing the wristband activity recorder could access a coloured, tapestry-like graph showing a 24-hour "lifestyle scene" over the experiment duration (Taniguchi *et al.* 2014). This enabled participants to identify at a glance any inconsistencies in their lifestyle habits such as working or sleeping hours. Organisers report that this experiment generated positive outcomes on health. In the group trialing both the life recorder and ICT scales, time spent exercising increased 34% while participants reported that sleep quality and daily vitality increased markedly. In addition, across all groups the majority of participants reported high satisfaction with the technologies and reported an increase in motivation to exercise.

The developer explained that this health data visualisation strategy was similar to the activities to monitor home energy consumption data. In both cases, ICT is used to influence user behaviour by providing visual "cues":

> The same goes for health. To assist people in leading a healthier lifestyle it is necessary to first of all quantify and then provide back visually that person's health condition. Therefore, the wearable life recorder for example aims to render visible that user's lifestyle pattern and then, by showing the duration or infrequency of sleep patterns, provide cues to improve this.

Participants could check the status of their health condition in real-time and as often as they desired via the Internet portal. This set of technologies therefore permitted vastly more frequent interaction with vital health indicators than an occasional hop on the scales or doctor's visit. This experiment was underpinned by grander ambitions of boosting the effectiveness of health care. The developer explained that the experiment organisers anticipated this data visualisation approach could be eventually commercialised to provide objective feedback data on lifestyle patterns to medical professionals, who traditionally, rely on anecdotal evidence from patients when assessing health conditions or illnesses origins. Although today's marketplace abounds with similar technologies like the Nike+ "Fuelband", the developer emphasised the cutting-edge and "epoch-making" nature of the prototype ICT devices since they were trialed a few years before similar technologies entered the market.

4.3.2. Education through information provision

Driven by assumptions that a more informed public would produce healthier choices and lifestyles, Kashiwanoha also fosters smart, healthy residents through various educational initiatives. In late 2014, the University of Tokyo enrolled the developer and 10 private corporations in the establishment of a physical headquarters for the smart city's focus on preventative health and longevity. This free-of-charge, walk-in educational facility called Ashita (see Figure 1) aims to increase citizen awareness and knowledge about preventative health and longevity strategies. Smartness in this approach thereby involves education and the provision of information to inform resident lifestyles. Although the facility targets all ages, around two-thirds of the roughly 2000 residents enrolled are over 60. Interestingly, Ashita's open-door policy allows free use by residents outside of the city, with some users travelling from neighbouring towns.

Education efforts are concentrated in three areas: walking, diet and socialising. The name "Ashita" reflects this triple focus by combining sounds from the Japanese words for "walk", "talk" and "eat". As shown by the focus on "talking", another defining feature of Ashita's distinctly human-centred approach concerns the resolve to advance preventative health and longevity by increasing socialisation opportunities. As noted above, researchers from the University of Tokyo established that elderly residents who have frequent social interactions and outings are less likely to develop dementia and walking disabilities. The socialisation strategy was also informed by feedback from various visualisation experiments where participants voiced the need for personal interaction to complement individual ICT devices and Internet-based communication with public health professionals.

There are various education activities at Ashita with different objectives. The facility provides visitors with demonstrations of exercise/strengthening techniques and explanations of products. Equipment manufacturers are allowed to freely display their devices to increase brand recognition and to spur sales. A second strategy includes the frequent hosting of events where invited health specialists deliver lectures on specific topics (e.g. alleviating joint pain, Nordic walking classes, etc.) and more informal coffee hours and musical performances that provide opportunities for socialisation. A third approach relies on data collection and assessment. Ashita boasts an array of sophisticated digital equipment supplied free-of-charge by medical equipment firms. The equipment allows visitors to monitor their physical condition through quantitative data on key vital indicators such as weight, BMI, basic metabolism, muscle weight and distribution, visceral fat, bone weight and artery health. The data are stored on membership cards to allow visitors to review their individual results over time. Anonymised data are also provided to equipment suppliers for commercial purposes. An interviewee admitted that while there were initial ambitions to create "big data" with commercial value, this is yet to be realised since only around a third or so of the 2000 enrolled members visit periodically. The centre staff are thus employing various marketing strategies to increase enrolment numbers, visitations and community awareness of the facility.

Far from a top-down approach, Ashita is run by a team of some 30 volunteers (all are local or nearby residents) and a few paid staff. Interestingly, none of the volunteers are health professionals. Instead, the centre provides them with basic training and empowers them to educate other residents to adopt pro-active and health-enhancing lifestyles. In connection to this collaborative learning focus, a volunteer staff member half-jokingly expressed hopes that visitors and volunteers would become "health evangelists" and spread healthy living habits throughout the community.

Beyond the health centre, the municipality has also seized the experimentation opportunities offered in the smart city to develop and test a transition to digital health records and an app-based health information and public education strategy. In late 2014 they partnered with the developer and several private IT firms to recruit 70 young mothers for a 30-day trial to assess the suitability of a cloud-based maternal and child health diary platform. This was intended to supplement traditional paper-based diaries to record basic health data such as height, weight and vaccination history. Paper-based child health diaries containing this information are a legal requirement for doctor visits in Japan. The smart diary fulfilled this basic function but also introduced dynamic temporal visualisations of the child's height and weight alongside healthy averages. The initial experiment harboured ambitions of seamlessly linking this function with health databases in the municipality that store medical information from child doctor visits. In addition, the diary was integrated with a vaccination appointment scheduling tool to minimise doctor visits and reduce time burdens on working parents.

More important, however, the smart diary was designed as an educational package to spur mutual learning around child health in couples and families. Accessible from smart phones or computers from either parent, it features for instance municipality produced video clips on correct bathing, holding and feeding techniques, and additionally, numerous text articles on various child health topics. Content of the digitally delivered health guidance is tailored to the age of the child. Interviewed health professionals from the municipality emphasised the importance of this tailored education in a society increasingly comprised of isolated nuclear families where young parents have few opportunities to learn from other child raising families. They also underscored that the transition to a smart diary addresses the growing societal need for digital and targeted health information that is easily accessible to career juggling couples.

The smart diary was received positively by users. This justified the launch of a full city-scale experiment in 2015 and the public launch of a formal municipal programme in 2016 (the first in Japan). While the pilot version was originally limited to privileged Kashiwanoha residents, it now serves the residents of the surrounding City of Kashiwa. However, the automatic data uploading platform used in the pilot study were not scaled up due to server maintenance costs and privacy concerns. Instead, health data must be manually uploaded by parents, significantly reducing the convenience that a "smart" diary promises. Reflecting on key outcomes, a municipality nurse emphasised how the transition from paper to smart diaries and digital education has facilitated enhanced communication around child health between residents and the municipality. A larger audience is now reachable than possible via traditional communication strategies such as pamphlets, posters and parenting classes. Reflecting this, interviewees shared feedback from the diary users that the relationship between young parents and the municipality now "felt intimate". Moreover, they also remarked that the digital educational materials produced by the municipality enjoy a particular level of trust due to their objective and tailored nature in a world abound with information on child health of varying reliability.

4.3.3. Enticement for behavioural change

Beyond information provision and technical experimentation, smart health initiatives in Kashiwanoha also involve efforts to entice behavioural change in residents via both intrinsic and extrinsic incentives. On the former, in several monitoring experiments, participants wearing individual ICT activity monitors or pedometers were linked into a digital community. A game-like, competitive ambience was fostered across the participant cohort by aggregating individual-level data for key indicators such as calories burned and steps walked or ran and then disclosing this via real-time graph visualisations on a cloud portal accessible from the Internet. For each indicator, participants could assess their personal performance against the cohort average. A rank was also generated for cumulative calories burned to show an individual's position relative to other "competitors". This approach was intended to spur ambitions of high achievement and create an intrinsic incentive whereby participants were rewarded with the personal satisfaction of exceeding the cohort average. By rendering visible the existence and efforts of other health crusaders through cohort-level health indicators, these experiments also fostered a sense of community and shared purpose to the traditionally individual affair of health and lifestyle management.

Smart technologies are also exploited to create extrinsic incentives. Smart city partners reasoned that "health data visualization alone is not sufficient for engaging the population segment that lacks interest in health. Some form of incentive is required". They developed a programme to provide performance-based financial rewards to residents in accord with steps walked or ran each month (as tracked by ICT pedometers). Rewards are collected via points that are redeemable for purchases in the developer's stores. Although they burden the cost of the programme, as they explained, "It is only by visiting our facilities that the user is able to redeem their points … and that's potentially good for our business". Thus, there is an explicit connection between public health and economic development. This approach was trialed extensively across various ICT experiments between 2012 and 2014. It now forms the basis of a Sukoyaka Links (meaning "healthy links") initiative, launched in late 2015 by the developer and a private medical technology firm. This Sukoyaka Links initiative involves the trial of three technological artefacts (see Figure 1 top right):

1. An ICT pedometer for recording distance steps walked/ran each day (1000 were distributed to residents and workers in Kashiwanoha free of charge by the developer);
2. An ICT card for receiving digital points based on the number of cumulative steps walked/ran (limited to a monthly maximum of 600 or the equivalent of JPY 600 or about US $5.50); and
3. A pedometer and point card scanning machine installed in the shopping centre for registering cumulative steps and calories burnt, generating visualisations of progress over time, and redeeming points.

The developer explained that learning accumulated from experiences in past experiments had significantly shaped the Sukoyaka Links initiative. First, by distributing free-of-charge pedometers to adult residents and workers in the smart city, the programme is a first attempt to engage the entire adult population of Kashiwanoha through ICT devices. This includes laggards not inclined to exercise and contrasts to earlier experiments, which mainly targeted smaller groups of motivated frontrunners. Second, relative to earlier experiments with ICT devices, a notable shift towards a streamlined and simplified thematic focus on walking occurred. User feedback collected by the organisers from earlier ICT experiments suggested that too many technological artefacts and types of information (e.g. different health indicators from ICT devices in conjunction with information from email exchanges with health professionals) risked overwhelming the user. Sukoyaka Links organisers thus reasoned that a simpler communication strategy would be more effective at fostering long-term engagement of the population around exercise.

> In the early years, our experimentation with technology demonstrated that technically a lot was possible … .But these experiments only focused on a small population segment with high interest in health. We have since realized that to attain a healthy city we must engage the "base of the pyramid" that is not usually interested in health. For that we require a population approach. We thus decided to narrow our thematic focus on simply "walking", which is easier to understand.

The third shift concerns a scaling-down of technological sophistication and dependency, which in essence, removes the need for a cloud server to store information. This eliminated the need for heavy investments in commercially vulnerable and immature technologies, since pedometers are readily available at low-cost on the market. However, despite much potential and careful design of the initiative, only 10% of the 1000 users receiving free pedometers regularly access the card scanning machine to redeem points. This suggests that the potential of intrinsic/extrinsic incentives created from technological devices to entice exercise only attracts a limited share of the population. This highlights the need for other population-wide approaches to target what the developer terms "the base of the pyramid".

4.4. Challenges in commercialisation, privacy and engaging residents

Private-sector-led experimentation with ICT and data visualisation technology was underpinned with lofty expectations that new and lucrative business development opportunities would emerge after the trials with residents. While experiments revealed a wide array of possibilities – both in regards to commercial applications and health-enhancing effects on users – none have reached full-scale development. The notable exception is the digital diary project, where investment and ownership were largely assumed by the municipality. Harsh business realities have betrayed the promise of profit for the private sector from smart health initiatives. Private-sector-led focus groups with residents trialing wearable ICT devices revealed a hesitancy to actually pay for such data services despite the fact that product satisfaction was high. Organisers noted that although the experiments succeeded in gaining the enthusiastic participation of residents by supplying technologies free of charge over brief intervals, a long-term commercial arrangement involving subscription fees would substantially dampen interest. As a compounding factor, an influx of competing technologies such as Nike Fuel bands and health apps on smart phones have overshadowed the novelty of the health devices trialed in Kashiwanoha. The developer also emphasised the unanticipated high costs of maintaining data servers and noted that their shared ambition to link real-time health data visualisations on an Internet portal alongside home energy visualisations was abandoned. Regarding financial burdens, the developer argued that public health should primarily be the responsibility of the municipality since their primary line of work concerned real estate development:

> On the issue of whose job it is to take care of health and welfare, it's the municipality that first comes to mind. But the municipality does not have the funds to invest heavily in ICT experiments and so on … As for us, we are not trying to profit from the health agenda so we too cannot afford to invest heavily.

They underscored a split-incentive issue where the private sector invested in the smart health strategy while the financial benefits such as reduced health care expenditures were reaped by the municipality. This undermines the incentive for the private sector to invest in smart health initiatives and raises interesting questions about who benefits from healthy residents and who should invest and take ownership of a smart health agenda.

In addition, the continuing inability for private firms to freely access individual citizen health data stored by the municipality (as part of the regular public health system) has significantly hampered the commercial development of the smart health agenda. Municipal datasets include individual-level health insurance expenditures, subsidised medicine intake, frequency of health visits and key health indicators obtained from health check-ups. Private-sector-led experiments and business plans over the years assumed that these data would be freely provided for commercial – albeit public health advancing – purposes. Instead, Kashiwanoha has become a graveyard for short-term experiments failing to overcome privacy concerns of residents and the reluctance of the municipality and associated medical advising groups to hand over datasets to private firms. This has hampered the creation of new ICT health services in the smart city since their raw fuel and commercial success depends on access to data – both that housed by the municipality and that collected from ICT devices. Data acquisition roadblocks have also severely undermined the ability of smart city actors to monitor and demonstrate the effect of health-advancing interventions on the population. The Ashita walk-in educational facility is one such example. Since smart city partners have been unable to verify the impact of this initiative on users through analysis of municipality health records, its effect on the health of users is uncertain, and its scientific legitimacy is undermined. There are signs, however, that this situation is slowly changing. Municipality datasets have been recently provided to one of the local universities, opening the door for potential future analysis of the health status of Kashiwanoha residents.

Finally, the smart health initiatives in Kashiwanoha reveal the inevitable difficulties of engaging a significant number of residents in the city. The point-based incentives programme was conceived as a means of engaging, on the one hand, the laggard population segment not inclined to exercise, and on the other hand, motivated frontrunners who might increase efforts in response to certain benefits. This frames residents as lazy *homo economicus* where interest in exercise can be unleashed via extrinsic and financial incentives. Interviews suggested that the developer's confidence that point-based incentives will continue to play a key role in engaging the "base of the pyramid" is unwavering. Such expectations mirror an increasing number of municipal initiatives in Japan that freely distribute pedometers to residents with linked point-based reward schemes. However, a university researcher suggested that this might raise social equity concerns since a point-based system rewards healthy individuals able to participate in the programme while neglecting those who are unwilling or unable to participate. In this situation, "the healthy who can walk get healthier" while those who cannot "lose out", they argued. They pointed out that while current financial incentives in the Sukoyaka Links initiative are funded by the developer, any future attempt to leverage municipal taxes (like other nationwide initiatives) would be problematic from an equity perspective, especially considering scientific uncertainty around the health effects of such schemes. These concerns point to a need for alternative approaches to engage the wider population in the pursuit of health.

As a step in this direction, the developer and private and academic partners have begun to embrace the emerging urban planning paradigm of an "active city", and in particular, a set of guidelines designed for New York City (City of New York 2010, 2013). To increase walkability, notable approaches advocated in the active city paradigm include mix-used development in close proximity to residences, street greenery and beatification to boost pleasure derived from walking, traffic calming features and interconnected footpaths with points of interest (street furniture, commercial/cultural facilities, etc.). A planning committee comprised of actors from the developer, local universities, the municipality, non-profits and industry is currently developing a Kashiwanoha vision for walkability and an associated set of guidelines. These will steer the remaining street planning in the smart city until completion in 2030. The emerging challenge for smart city planners concerns the question of how to link the historically established urban engineering approach that influences population lifestyles through built environment design with exercise incentives and behaviour signals created from ICT devices.

5. Discussion

Kashiwanoha's experiences suggest that smart strategies have the potential to transform health and well-being from an individual pursuit to a collective and social endeavour. Early experiments demonstrated that wearable ICT devices can collect individual-level data and then contribute to community building by allowing participants to compare individual performance with others. This allows for scrutiny of individual efforts (steps walked, calories burned, etc.) from a community perspective, connecting the individual to a wider network of fellow residents sharing a common goal of improving health and well-being. Similarly, the digital maternal and child health diary project explicitly seeks to promote joint learning across couples and families by allowing multiple users to access common educational materials. It also goes a step further by connecting residents and municipality health advisors through an app-based email function. User feedback suggests that this lowers communication barriers between the municipality and residents, promoting a sense of comfort for young parents in knowing that a nurse is never more than a button click away.

Complementing the virtual community, the walk-in Ashita centre is an interesting beacon of community building around shared values of preventative health. It emphasises health maintenance in seniors through socialising and offers numerous social gatherings and group lessons. As explained earlier, the centre is run by resident volunteers, creating an arena of collaborative learning and education among non-experts. The initiative suggests the possibility of a bottom approach to smart city governance that is driven by citizen empowerment (Capdevila and Zarlenga 2015, Vanolo 2016, March and Ribera-Fumaz 2016). Equally, by enhancing citizen knowledge around preventative health via informal education, the Ashita centre has the potential to reorient smartness towards people and "a knowledge-intensive rather than a technology-intensive vision of cities and their development" (Söderström 2016, p. 65). Citizen learning and capacity building in the Ashita facility contrast to the passive role of residents in ICT visualisation experiments, which was largely restricted to providing feedback to industry. This strategy of fostering smart and empowered residents appears particularly novel considering that researchers frequently criticize the smart city agenda – both inside and outside Japan – for not delivering on promises to facilitate more participatory governance (Gabrys 2014, Glasmeier and Christopherson 2015, Hollands 2015, Granier and Kudo 2016).

The contrasting strategies exploited to pursue a smart health agenda merit reflection. Guided by iterative learning and accumulated experiences, the developer facilitated this process by archiving experiences and playing a long-term and central role in all projects. One strategy to realise a smart, healthy city relies on sophisticated technology, intensive data collection and visual feedback. This involves a commercially risky technology push-approach from private actors, a focus on smart devices, and an initial targeting of motivated frontrunners. The approach represents a novel but direct application of the data visualisation and feedback paradigm used to influence demand side energy consumption in residential buildings (Gölz and Hahnel 2016, Timm and Deal 2016). This suggests that resident lifestyles in a smart city can be mined to produce more than conventional sustainability indicators such as building-level energy consumption. Visualisation experiments also demonstrated the power of these data, with positive health impacts resulting from participant interactions with devices and feedback technologies. Ambitions to couple data collected from individual-level ICT devices with municipality housed health and expenditure records also promises a novel way to monitor resident well-being and verify the impacts of public health interventions.

However, data acquisition challenges provide a sobering reminder to other aspiring smart cities that data-driven approaches to health management risk bumping against social norms. These principally concern the individual information that citizens and municipalities are prepared to share with market actors due to privacy and security concerns. Also, prompted by harsh business realities in ambitions to commercialise several technologies, Kashiwanoha actors were prompted to pursue alternative avenues for smart health. These consist of more people-centred and less data-dependent approaches including education, incentive building for behavioural change and a strong focus on creating socialisation opportunities. These softer, low-tech approaches offer a notable shift from data-driven, high-tech narratives about smart cities (Townsend 2014, March and Ribera-Fumaz 2016, March in press).

Lastly, we highlight findings that several experiments and initiatives in Kashiwanoha were opened to residents and commuting workers from outside the smart city. For example, the walk-in centre Ashita allows free-of-charge access to preventative health check-ups, education and socialising events to all citizens – regardless of area of residence. Similarly, although first trialed exclusively in the smart city, the electronic child health diary initiative has now been upscaled to provide free-of-charge access to all interested parents in the surrounding and larger City of Kashiwa. Additionally, several experiments including the Sukoyaka Links initiative targeted both local residents and commuting workers. By addressing a population beyond the exclusive and economically fortunate residents of Kashiwanoha, these open-door policies provide a direct way to address equity concerns. Many smart cities risk exasperating social polarisation by concentrating benefits (e.g. healthier cleaner environments, advanced infrastructure and services, economic vitality, etc.) in specific areas while excluding less privileged populations from participation (Caprotti et al. 2015, Hollands 2015). This points to the argument that even with the seemingly widespread issue of public health, there is a need to consider who benefits and who does not benefit from these activities.

6. Conclusion

Using an illustrative case of smart health and well-being strategies in a smart city in Japan, this study set out to build understanding into how smart city agendas can be stretched beyond narrow aims of environmental protection and economic development to encompass social issues and lifestyles. The findings contribute to an emerging discourse on the possibilities for smart cities to utilise digital technologies and smart planning strategies to improve the human dimensions of urban life. Active pursuit of improved public health in Kashiwanoha has become a key part of the city's identity. It positions itself as a "campus" for the local universities and boasts its international and health-enhancing profile that is informed by academic knowledge. Our examination of a combination of data visualisation experiments, public education and differing incentives for behavioural change illustrate multiple ways that a potentially generic agenda of smart health was tailored to specific needs and a diversity of residents. These strategies respond to the urgent need to address the mounting challenge of an ageing population in Japan and its increasing burdens on public health care. This specific and endogenous social problem heavily defined Kashiwanoha's smart city agenda focus on preventative health and longevity, then prompting a search for specific ICT solutions. The case study suggests a novel approach to smart urban development. Digital technologies are framed not as an end in themselves, but rather, as tools to tackle social issues and improve citizen livelihoods. Additionally, by tackling the issue of health and longevity, smart city partners in Kashiwanoha have developed various strategies to intervene on the lifestyles of residents. This offers a novel illustration of how the smart city agenda can potentially generate important opportunities to apply smart technologies to the goal of enhancing the lives and well-being of residents.

The stretched smart city agenda in Kashiwanoha holds significant implications on the roles and motivations of the various urban stakeholders involved in pursuing greater public health and well-being. The developer's leadership in the smart health agenda cannot be explained by altruism or corporate social responsibility. Instead, they recognised that the smart health objective would create brand differentiation in a competitive real estate market, and increase the commercial attractiveness of their assets. The municipality too played a leading role in experimenting with digital technologies and developing new services for residents. They hold a clear interest in reducing public health care expenditures. This motivation prompted a shift beyond traditional roles of simply providing basic infrastructure and services in return for taxes to an active digital educator role to assist residents in making informed choices about health and well-being. Smart health thereby offers an opportunity for local authorities to reimagine public service delivery while relying on technical expertise from the private sector. This has the potential to reinforce the local government's role and increase their relevance and legitimacy as providers of collective urban services. This has the potential to mitigate concerns about the neoliberal tenor of smart urban development and the privatisation of collective services (Glasmeier and Christopherson 2015, March in press). Findings also showed how residents were actively involved in bottom-up and joint learning initiatives such as the walk-in preventative health centre Ashita and intrinsic/extrinsic rewards programmes. The experiences in Kashiwanoha suggest a more fluid understanding of local governance as an endeavour shared by multiple stakeholders with contrasting yet compatible motives.

As we have shown, the smart health agenda in Kashiwanoha complements and extends existing aims to advance environmental protection and business development. It demonstrates how smartness can serve as a thread binding the three sides of the sustainability triangle (i.e. environment, economy and society). It also provides convincing evidence that the notion of a smart city harbours potential to be stretched to encompass wider social considerations such as "the intangible entity of a citizen's wellbeing" (Glasmeier and Nebiolo 2016, p. 9). Moreover, findings hint at the wider opportunity for the smart urban development agenda to address an even wider diversity of social issues such as crime, poverty, education and social cohesion (Glasmeier and Christopherson 2015, Goodspeed 2015, Bibri and Krogstie 2017). Finally, although the empirical evidence from this case study is far from definitive, it also points to the importance of customising universal smart city principles, technologies and development strategies to address specific local circumstances and needs.

Acknowledgements

The authors express their sincere gratitude to the interviewees who kindly donated their time to support this study.

Disclosure statement

No potential conflict of interest was reported by the authors.

ORCID

Gregory Trencher http://orcid.org/0000-0001-8130-9146
Andrew Karvonen http://orcid.org/0000-0002-0688-9547

References

Albino, V., Berardi, U., and Dangelico, R.M., 2015. Smart cities: definitions, dimensions, performance, and initiatives. *Journal of Urban Technology*, 22, 3–21.
Alizadeh, T., 2017. An investigation of IBM's smarter cites challenge: what do participating cities want? *Cities*, 63, 70–80.
Andreassen, H.K., Kjekshus, L.E., and Tjora, A., 2015. Survival of the project: a case study of ICT innovation in health care. *Social Science & Medicine*, 132, 62–69.
Arrizabalaga, S., Seravalli, A., and Zubizarreta, I., 2016. Smart city concept: what it is and what it should be. *Journal of Urban Planning and Development*, 142 (1), 04015005.
Bakıcı, T., Almirall, E., and Wareham, J., 2013. A smart city initiative: the case of Barcelona. *Journal of the Knowledge Economy*, 4, 135–148.
Bibri, S.E., and Krogstie, J., 2017. On the social shaping dimensions of smart sustainable cities: a study in science, technology, and society. *Sustainable Cities and Society*, 29, 219–246.
Capdevila, I., and Zarlenga M.I., 2015. Smart city or smart citizens? The Barcelona case. *Journal of Strategy and Management*, 8, 266–282.
Caprotti, F., Springer, C., and Harmer, N., 2015. "Eco" for whom? Envisioning eco-urbanism in the Sino-Singapore Tianjin Eco-city, China. *International Journal of Urban and Regional Research*, 39, 495–517.
City of New York, 2010. *Active design guidelines: promoting physical activity and health in design*. New York: City of New York.
City of New York, 2013. *Active design: shaping the sidewalk experience*. New York: City of New York.
DeWit, A., 2013. Japan's rollout of smart cities: what role for the citizens? *The Asia-Pacific Journal*, 11, 1–12.

Gabrys, J., 2014. Programming environments: environmentality and citizen sensing in the smart city. *Environment and Planning D: Society and Space*, 32, 30–48.

Giles-Corti, B., *et al.*, 2016. City planning and population health: a global challenge. *The Lancet*, 388, 2912–2924.

Glasmeier, A. and Christopherson, S., 2015. Thinking about smart cities. *Cambridge Journal of Regions, Economy and Society*, 8, 3–12.

Glasmeier, A. and Nebiolo, M., 2016. Thinking about smart cities: The travels of a policy idea that promises a great deal, but so far has delivered modest results. *Sustainability*, 8, 1122.

Gölz, S. and Hahnel, U.J.J., 2016. What motivates people to use energy feedback systems? A multiple goal approach to predict long-term usage behaviour in daily life. *Energy Research & Social Science*, 21, 155–166.

Goodspeed, R., 2015. Smart cities: moving beyond urban cybernetics to tackle wicked problems. *Cambridge Journal of Regions, Economy and Society*, 8, 79–92.

Granier, B., and Kudo, H., 2016. How are citizens involved in smart cities? analysing citizen participation in Japanese "smart communities". *Information Polity*, 21, 61–76.

Haarstad, H., 2016. Constructing the sustainable city: examining the role of sustainability in the "smart city" discourse. *Journal of Environmental Policy & Planning*, 1–15. doi:10.1080/1523908X.2016.1245610.

Haluza, D. and Jungwirth, D., 2015. ICT and the future of health care: aspects of health promotion. *International Journal of Medical Informatics*, 84, 48–57.

Hasan, H. and Linger, H., 2016. Enhancing the wellbeing of the elderly: social use of digital technologies in aged care. *Educational Gerontology*, 42, 749–757.

Hielkema, H. and Hongisto, P., 2013. Developing the Helsinki smart city: the role of competitions for open data applications. *Journal of the Knowledge Economy*, 4, 190–204.

Hodson, M. and Marvin, S., 2017. Intensifying or transforming sustainable cities? Fragmented logics of urban environmentalism. *Local Environment*, 1–15. doi:10.1080/13549839.2017.1306498.

Hollands, R.G., 2015. Critical interventions into the corporate smart city. *Cambridge Journal of Regions, Economy and Society*, 8, 61–77.

Karvonen, A., 2011. *Politics of urban runoff: nature, technology, and the sustainable city*. London: The MIT Press.

Kijou, N. and Rure, K. 2014. *Smart cities: observations on issues regarding commercialisation of verification experiments (in Japanese)*. Tokyo: EY Institute.

Koch, S., 2010. Healthy ageing supported by technology – a cross-disciplinary research challenge. *Informatics for Health and Social Care*, 35, 81–91.

Kono, N., Suwa, A., and Ahmad, S., 2016. Smart cities in Japan and their application in developing countries. *In*: J. Jupesta and T. Wakiyama, eds. *Low carbon urban infrastructure investment in Asian cities, cities and the global politics of the environment*. London: Palgrave, 95–122.

Kurata, N., *et al.*, 2013. Campus planning for promoting quality of life in the community. *In*: A. Konig, ed. *Regenerative sustainable development of universities and cities: the role of living laboratories*. Northampton, MA: Edward Elgar, 236–253.

Lee, J.H., Hancock, M.G., and Hu, M.-C., 2014. Towards an effective framework for building smart cities: lessons from Seoul and San Francisco. *Technological Forecasting and Social Change*, 89, 80–99.

Leeds City Council, & U. o. Leeds., 2015. *Cities as places of wellbeing: smart cities health & wellbeing*. Leeds: Leeds City Council.

Mah, D.N.-y., *et al.*, 2013. The role of the state in sustainable energy transitions: a case study of large smart grid demonstration projects in Japan. *Energy Policy*, 63, 726–737.

March, H., in press. The smart city and other ICT-led techno-imaginaries: any room for dialogue with degrowth? *Journal of Cleaner Production*.

March, H. and Ribera-Fumaz, R., 2016. Smart contradictions: the politics of making Barcelona a self-sufficient city. *European Urban and Regional Studies*, 23, 816–830.

Meijer, A. and Bolívar, M.P.R., 2016. Governing the smart city: a review of the literature on smart urban governance. *International Review of Administrative Sciences*, 82, 392–408.

Melosi, M.V., 2000. *The sanitary city: urban infrastructure in America from colonial times to the present*. Baltimore, MD: John Hopkins University Press.

Mitsui Fudosan, 2017. *Kashiwa-no-ha smart city*. Mitsui Fudosan. Available from: http://www.kashiwanoha-smartcity.com/en/concept/whatssmartcity.html [Accessed 25 July 2017].

Muramatsu, N. and Akiyama, H., 2011. Japan: super-aging society preparing for the future. *The Gerontologist*, 51, 425–432.

Nyberg, R. and Yarime, M., 2017. Assembling a field into place: smart-city development in Japan. *Research in the Sociology of Organizations*, 50, 253–279.

Obi, T., Ishmatova, D., and Iwasaki, N., 2013. Promoting ICT innovations for the ageing population in Japan. *International Journal of Medical Informatics*, 82, e47–e62.

Pincetl, S., 2010. From the sanitary city to the sustainable city: challenges to institutionalising biogenic (nature's services) infrastructure. *Local Environment*, 15, 43–58.

Ramaswami, A., *et al.*, 2016. Meta-principles for developing smart, sustainable, and healthy cities. *Science*, 352, 940–943.

Sallis, J. F., *et al.*, 2016. Use of science to guide city planning policy and practice: how to achieve healthy and sustainable future cities. *The Lancet*, 388, 2936–2947.

Saujot, M., and Erard, T., 2015. *Smart city innovations for sustainable cities? An analysis based on data challenges.* Paris: Institut du développement durable et des relations internationales.

Schuurman, D., *et al.*, 2012. Smart ideas for smart cities: investigating crowdsourcing for generating and selecting ideas for ICT innovation in a city context. *Journal of Theoretical and Applied Electronic Commerce Research*, 7, 49–62.

Shirahase, S., 2015. Demography as destiny: falling birthrates and the allure of a blended society. *In*: F. Baldwin and A. Allison, eds. *Japan: the precarious future*. New York: New York University Press, 11–35.

Söderström, O., 2016. From a technology intensive to a knowledge intensive smart urbanism. *In*: J. Stollmann, K. Wolf, A. Brück, S. Frank, A. Million, P. Misselwitz, J. Schlaack, and C. Schröder, eds. *Beware of smart people! Redefining the smart city paradigm towards inclusive urbanism*. Berlin: Universitätsverlag der TU Berlin, 63–69.

Stollmann, J., *et al.*, 2015. *Beware of smart people! Redefining the smart city paradigm towards inclusive urbanism*. Berlin: Universitätsverlag der TU Berlin.

Taniguchi, S., *et al.*, 2014. The effect of health data visualization to promote healthy behavior: a summary of the Kashiwa-no-ha Smart City Project. *In*: *AAAI spring symposium* 2014. Palo Alto, CA: Stanford University/Association for the Advancement of Artificial Intelligence, 100–101.

Thomas, F., *et al.*, 2014. Extended impacts of climate change on health and wellbeing. *Environmental Science & Policy*, 44, 271–278.

Timm, S.N., and Deal, B.M., 2016. Effective or ephemeral? The role of energy information dashboards in changing occupant energy behaviors. *Energy Research & Social Science*, 19, 11–20.

Townsend, A.M. 2014. *Smart cities: big data, civic hackers, and the quest for a new utopia*. New York: W. W. Norton.

Tranos, E., and Gertner, D., 2012. Smart networked cities? *Innovation: The European Journal of Social Science Research*, 25, 175–190.

Trencher, G., Terada, T., and Yarime, M., 2015. Student participation in the co-creation of knowledge and social experiments for advancing sustainability: experiences from the university of Tokyo. *Current Opinion in Environmental Sustainability*, 16, 56–63.

UDCK. 2017. *Vision of UDCK*. Urban Design Centre Kashiwa. Available from: http://www.udck.jp/en/about/002443.html [Accessed 25 July 2017].

Vanolo, A., 2016. Is there anybody out there? The place and role of citizens in tomorrow's smart cities. *Futures*, 82, 26–36.

Viitanen, J. and Kingston, R., 2014. Smart cities and green growth: outsourcing democratic and environmental resilience to the global technology sector. *Environment and Planning A*, 46, 803–819.

Yarime, M. and Karlsson, M., in press. Understanding the innovation system of smart cities: the case of Japan and implications for public policy and institutional design. *In*: J. Niosi, ed. *Innovation policy, systems and management*. Cambridge: Cambridge University Press.

Urban sharing in smart cities: the cases of Berlin and London

Lucie Zvolska ⓘ, Matthias Lehner, Yuliya Voytenko Palgan, Oksana Mont and Andrius Plepys

ABSTRACT

Addressing urban sustainability challenges requires changes in the way systems of provision and services are designed, organised and delivered. In this context, two promising phenomena have gained interest from the academia, the public sector and the media: "smart cities" and "urban sharing". Smart cities rely on the extensive use of information and communications technology (ICT) to increase efficiencies in urban areas, while urban sharing builds on the collaborative use of idling resources enabled by ICT in densely populated cities. The concepts have many similar features and share common goals, yet cities with smart city agendas often fail to take a stance on urban sharing. Thus, its potentials are going largely unnoticed by local governments. This article addresses this issue by exploring cases of London and Berlin – two ICT-dense cities with clearly articulated smart city agendas and an abundance of sharing platforms. Drawing on urban governance literature, we develop a conceptual framework that specifies the roles that cities assume when governing urban sharing: city as regulator, city as provider, city as enabler and city as consumer. We find that both cities indirectly support urban sharing through smart agenda programmes, which aim to facilitate ICT-enabled technical innovation and emergence of start-ups. However, programmes, strategies, support schemes and regulations aimed directly at urban sharing initiatives are few. We also find that Berlin is sceptical towards urban sharing organisations, while London took more of a collaborative approach. Implications for policy-makers are discussed in the end.

Introduction

In 1950, 30% of the world's population was living in cities. By 2050, this figure is expected to rise to 66% (UN 2015). Such rapid urbanisation bears with it a number of sustainability challenges. Urban population is currently contributing to as much as 80% of overall energy consumption, 75% of carbon emissions and 75% of global natural resource consumption (UNEP 2013). At the same time, over 80% of the global gross domestic product is generated in cities (Grübler and Fisk 2013). Thus, cities offer many socio-economic benefits and have become centres of research and education (König and Evans 2013; Trencher et al. 2014), catalysing a change from the unsustainable status quo. In recent years, two concepts that promise to tackle urban sustainability challenges have emerged: smart cities and the sharing economy.

Smart cities and information and communications technology (ICT)-enabled urban sharing[1] boast somewhat similar values and all-encompassing goals and exhibit a potential to contribute to digitally enabled green urbanism (Hollands 2008; McLaren and Agyeman 2015). Gori, Parcu, and Stasi (2015) presented several common features of the sharing economy and smart city concepts. They framed the concepts as innovations led by citizens' and consumers' needs, functioning in well-defined communities, whose aim is to share resources, be it material resources, skills, time or data, through ICTs.

The European Union defines the concept of smart cities as the utilisation of "scalable solutions that take advantage of information and communications technology (ICT) to increase efficiencies, reduce costs, and enhance quality of life" (EC 2013, 5). It has also been described as an innovative concept, which is expected to provide solutions to many societal challenges. While it lacks a definitional precision, it is often associated with the utilisation of solutions based on ICT that increase efficiencies and bring together public actors, citizens and private companies (Höjer and Wangel, 2015). However, smart cities are criticised for being corporatised (Hollands 2015). Some researchers are also questioning its true potential to improve economic, environmental and social conditions for urban dwellers. Furthermore, Vanolo (2014) demonstrates on a case from Italy that the smart city idea creates an uncritical consensus that limits cities' planning approaches to a single concept. Hollands (2015) further argues that if cities are to address social problems, they need to shift from the "technologically driven, corporately controlled, heavily marketed" smart city to citizen-led, participatory governance. One way to ensure that smart cities live up to their expectations is to support small-scale, bottom-up initiatives.

Similarly, Dyer, Gleeson, and Grey (2017) argue that the abundance of ICT in cities alone is not necessarily sufficient to deliver more inclusive governance processes known as "collaborative urbanism", which builds on the engagement of citizens and local actors in governing the city. Therefore, there is still a need for "a culture of Internet inspired citizen participation" to become mainstream (Gleeson and Dyer 2017). One way to ensure that smart cities live up to their expectations is for the city governments to support small-scale, bottom-up initiatives (Hollands 2015), as well as to engage in co-creation processes with local actors through partnerships, experimental arenas (Bulkeley and Castán Broto 2013) and urban living laboratories (cf. Evans and Karvonen 2014; Menny, Voytenko Palgan, and Mccormick 2018; Voytenko et al. 2016).

Within the wide landscape of smart technologies, initiatives and solutions that emerge in cities, the ICT-enabled urban sharing is a particularly promising example of smart innovation that offers a novel way of interaction and resource use between urban actors. Urban sharing encompasses a wide array of communal and commercial urban sharing organisations (USOs) that employ ICT to reduce transaction costs and make sharing of resources among peers easily accessible. One of the first urban sharing programmes was bike sharing schemes, such as Barcleys Cycle Hire. Nowadays, a wide array of resources is being shared in cities; from garden tools and food to apartments and cars. A large population density and an accumulation of resources in cities have resulted in a rapid growth of USOs in many urban areas. Much of the attention of the media, city officials and academics have been on larger platforms, which have been successful in scaling up and even disrupting the incumbent industries (Guttentag 2015; Laurell and Sandström 2016). However, their potential to deliver sustainable solutions in cities remains to be realised (Cohen 2016; Schor 2014). Similarly to the smart city concept, it is bottom-up, small-scale organisations that have been praised for their potential to contribute to sustainable cities built on social justice, democratic collaboration and trust, but they either have difficulties multiplying or scaling up (Cohen 2016; Schor 2014; Sundararajan 2016) or lack the intention to do so.

City governments around the world are increasingly adopting policies to regulate some forms of USOs, and a number of municipalities are even taking legal action against them due to breached employment, zoning, and health and safety laws and licensing rules (Orsi 2013). Consequently, official reports and academic texts have emerged with suggestions on how cities should regulate these platforms (cf. Katz 2015; Sundararajan 2016; Wosskow 2014). In general, they suggest ways to promote innovation while also protecting USOs' consumers, but often neglect to take into consideration the sustainability merits of various urban sharing models.

City governments have the power to encourage or discourage certain types of sharing, but there is a lack of understanding of how cities govern USOs, and what role smart city agendas play in their dissemination (Agyeman, McLaren, and Schaefer-Borrego 2013; Cohen and Kietzmann 2014). This article seeks to address this gap by drawing on urban governance literature, specifically on its stream that deals with the role of cities in tackling sustainability challenges (Bulkeley and Kern 2006; Bulkeley et al. 2010; Evans and Karvonen 2010; Kern and Alber 2008). This literature emerged from the realisation that cities provide a fruitful ground for technical, social and political innovation (Cattacin and Zimmer 2016; Oecd 2008). In particular, we build on frameworks that describe four modes of urban climate governance (Bulkeley and Kern 2006; Kern and Alber 2008): self-governing, governing by authority, governing by provision and governing through enabling. Although these earlier versions of the urban climate governance frameworks tended to side-track participatory governance and the notion of social justice, we find them helpful to understand not only how municipalities seek to address the challenge of sustainable development, but also how they intervene in socio-technical processes by engaging with different types of innovations, including smart cities and urban sharing. In this article, we analyse the modes of urban governance in relation to urban sharing in two case cities: Berlin and London. These two ICT-dense cities have clearly articulated smart city agendas, a vibrant urban sharing landscape with the different prominence of for-profit and non-profit sharing organisations, and distinct ways of governance and engagement with urban sharing. While London has for a long time been a test-bed for a great number of business model innovation projects and sharing start-ups, Berlin has been known for hosting grassroots sharing organisations and has an ambition to become the sharing capital of Europe.

The paper is guided by the following research question: *How do city governments of Berlin and London govern ICT-enabled urban sharing organisations?*

Section 2 provides an overview of urban governance literature and describes the modes of urban governance in more detail. Section 3 deals with the methodological approach employed in this paper. It is followed by a case analysis of Berlin and London (section 4), discussion (section 5) and conclusions (section 6).

Conceptual framework

In seeking to capture the current political processes unfolding in European cities, scholars have moved from the term "government" to "governance" (Bulkeley and Kern 2006; Cattacin and Zimmer 2016). Governance describes "non-hierarchical modes of coordination, steering and decision-making" (Cattacin and Zimmer 2016), which apart from the representatives of formal government structures include an array of urban actors from private and public domains. Governance processes are thus pictured as regulations through "networks of agents" (Khan, 2013; Powell 1990). When analysing how governance of climate change is exercised at the municipal level in the U.K. and Germany (Bulkeley and Kern 2006) and in multi-level systems across the globe (Kern and Alber 2008), four distinct governance modes have been identified: governing by authority, governing by provision, governing through enabling and self-governing.

Governing by authority is the most traditional process of governance that builds on formal planning, control and regulation, and relies on legal sanctions by the jurisdiction to assure implementation. Kern and Alber (2008) label this mode more narrowly as "governing by regulation", which encompasses the formal authority of the city to steer by laws and policies. While being the most traditional governance mode, it is not the most popular one in climate change governance, as municipalities seek to avoid resistance and possible conflicts as a result of exercising their authority (Bulkeley and Kern 2006; Kern and Alber 2008).

Governing by provision focuses on the delivery of particular forms of services and resources through infrastructure and financial policy (Kern and Alber 2008). It is often identical to the role of the municipality as a shareholder in a local utility company for the provision of energy, transport, water or waste management services. It appears to be a common practice in more socially oriented countries with high taxes and the resulting high municipal budgets. This mode has not been found very prominent in relation to climate change governance because energy markets are experiencing increased liberalisation, and the ownership of utilities is being transferred to business actors (Bulkeley and Kern 2006).

Governing through enabling is a "softer" and less resource-intensive approach to municipal governance. It is reflected in the actions of municipalities that coordinate and facilitate partnerships with private actors and encourage community engagement. It relies on persuasion, argument and incentives (Bulkeley and Kern 2006). This mode has various dimensions, including public education, awareness campaigns and promotional activities by municipalities, developing external ties (e.g. co-operation by the city with other actors) and facilitating co-operation between stakeholders (e.g. the establishment of public–private partnerships for the provision of services and infrastructure) (Kern and Alber 2008). Khan (2013) refers to this governance role as "network governance", in which "the municipality is a facilitator rather than commander and implementer" (133). Co-operation and trust are essential for this way of governing (Khan 2013).

Self-governing refers to the municipality governing its own activities, for example, by improving energy efficiency in municipally owned buildings (Kern and Alber 2008). It relies on reorganisation, institutional innovation and strategic investments. It reflects two roles of a municipality: as a consumer through public procurement, and as a role model by visualising the feasibility of desirable solutions to other urban actors. Through self-governing and enabling, municipalities had the most discretion and decision-making power (Kern and Alber 2008). At the same time, self-governed activities only resulted in marginal contributions to urban climate change mitigation due to their small scale.

Drawing on the urban governance modes, we conceptualise four roles city governments may assume when working with a broad range of issues, either by supporting and promoting or by inhibiting them: city as regulator, city as provider, city as enabler and city as consumer (Figure 1). We suggest that these roles are relevant in addressing how municipalities engage with urban sharing.

"City as regulator" employs a range of regulatory mechanisms that include laws, taxes, bans, policies and other formal documents that regulate the establishment and operation of urban sharing initiatives. "City as provider" offers financial (i.e. "city as investor") and infrastructural (i.e. "city as host") support to USOs. Municipal funding programmes can be used by USOs to invest in their

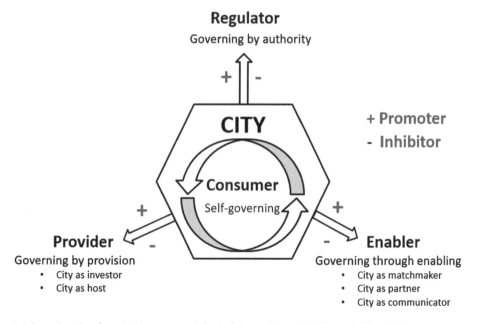

Figure 1. Roles and modes of municipal governance (after Bulkeley and Kern 2006; Kern and Alber 2008).

core activities (e.g. technology, infrastructure) or in their development work (e.g. personnel, research, communication and education) (Kern and Alber 2008). "City as enabler" may facilitate collaboration among USOs (i.e. "city as matchmaker") or creation of partnerships with municipal actors (i.e. "city as partner"). The city may organise competitions, awards and voluntary certification schemes to recognise the best sharing practices. It may also get engaged in disseminating the best urban sharing practices and in marketing them to different stakeholders (i.e. "city as communicator"). The role of "city as consumer" can be exemplified by municipalities adopting urban sharing practices in their own operations, such as procurement.

The city government can employ any of the four roles and combine them to varying degrees when dealing with any one issue (Bulkeley and Kern 2006). In addition, the roles can be played out as either promoting or inhibiting the emergence and operation of USOs. Both promotion and inhibition could be explicit, e.g. when cities incentivise or ban a certain activity, or subtle, e.g. when cities choose to support an alternative activity, which results in side-tracking or de-prioritising other USOs. This gives city governments a high degree of freedom to adopt the most suitable portfolio of methods and ways of working with urban sharing. At the same time, the course of action by a city government is affected by how the government is organised, by the local and national contexts, and the socio-technical configurations of the city, such as infrastructure, culture and economy (Hodson and Marvin 2010). Municipalities are also constrained in their actions by a world of multi-level governance where local decision-makers are dependent on both higher political levels and other actors in the society (Kern and Alber 2008; Khan 2013). Therefore, the city's capacity to deploy different modes of governance differs considerably (Bulkeley and Kern 2006). Following this framework, we conduct a comparative analysis of how municipalities engage with ICT-enabled sharing employing multiple governance roles.

In the context of this study, it is necessary to be aware of the administrative differences between Berlin and London. Berlin is located in the Federal Republic of Germany. It is one of three "city-states" in Germany with broad executive rights. This implies that it has executive power on two out of three executive levels within Germany; the municipality and the region.

The local government in London takes place in two tiers. The upper tier consists of The Greater London Authority (GLA) led by the Mayor of London who provides city-wide leadership, and the 25-member London Assembly who review the Mayor's policies, actions, strategies and budget plans. In addition to promoting economic, social and environmental development in the Greater London area, the GLA is responsible for policies on transport, buildings and land use (GLA 2016). Mayoral policies may influence daily operations of USOs, as was the case of changing private hire regulations or lobbying the national parliament to enforce stricter rules for short-term rentals (BBC 2016; Sullivan 2016). The GLA shares local government powers with the lower tier, which consists of the City of London Corporation and 32 boroughs that provide a large variety of services, including education, housing, planning, social services, environmental health or collection of council tax.

Berlin is located in Germany, a federal state. This system gives broad executive rights to regional (Bundesland) and local (Stadt, Gemeinde) authorities. Berlin, being the capital of the Federal Republic of Germany, is also one of only three so-called city-states in Germany. This implies that the city of Berlin has legislative power on two out of three legislative levels within Germany; the municipality and the region. The city of Berlin is further divided into 12 so-called Bezirke. These sub-units of the municipal government are executive bodies of the municipality. Their function is to institute the federal, regional and municipal legislation. Due to the federal nature of Germany, the elected government of Berlin has legislative powers in areas such as transport, rent, tourism and trade, and thus a significant influence over the nature of the sharing economy within Berlin.

Methodology

This research uses a deductive research strategy in which literature analysis is combined with case studies of Berlin and London as two cities promoting smart and sustainable agendas.

First, a *literature analysis* was conducted that identified existing knowledge about the roles cities can play in governing different types of developments (Bulkeley and Kern 2006; Gibson, Robinson, and Cain 2015; Kern and Alber 2009; Kronsell and Mukhtar Landgren 2017). The literature analysis included a review of data from multiple sources including academic publications, scientific and business reports, grey literature, periodicals and online sources, including official webpages for London and Berlin, and organisations dealing with smart agendas in these cities. We make use of online databases, such as Web of Science, SCOPUS, Google Scholar, EBSCO Host, LUBRIS and Research Gate. Keywords for searching in secondary sources included "sharing economy", "smart cities", "sharing city", "urban sharing" and "governance".

Second, based on the literature analysis, an *analytical framework* was developed that was used for data collection and data structuring. It was first employed for developing two interview guides and then for categorising and presenting the findings. One interview guide was designed for sharing organisations and one for city governments. Each guide comprised eight open-ended questions (Appendix 1).

The interviews were semi-structured, giving a possibility to the interviewed societal actors, both USOs and authorities, to put forward their insights and to elaborate on interpretations. In Berlin, the interviews were conducted in German, and in London in English.

Third, case studies from two cities, London and Berlin, were developed with data collected via a comprehensive literature analysis of city initiatives within smart and sharing agendas, complemented with *24 interviews* with 26 relevant stakeholders representing city governments, sharing organisations and stakeholders collected through chain sampling (Appendix 2).

Ten interviews were conducted in London and 14 in Berlin (please see Appendix 2 for an anonymised list of all interviewees). The interviews were conducted by phone and lasted between 30 and 90 minutes. One interview was conducted via email with a public officer from the GLA. The interviews were recorded and thematically transcribed. The different roles of cities presented in the analytical framework were tested either "in vivo", i.e. through direct words and framings used by the interviewees, or by summarising the concepts discussed by them. Detailed findings from Berlin and London are presented below.

Findings and analysis

Berlin declared itself a "smart city" and adopted the Smart City Berlin Strategy in 2015. It has since dedicated resources to the development of the smart city agenda, for example, by creating a cross-departmental unit within the municipality. The sharing idea has also received some attention and resources from the city of Berlin, particularly in late 2014 and early 2015, when the business development unit of Berlin organised meetings, workshops and commissioned a report on the potential of urban sharing. However, unlike the smart city agenda, the interest in sharing has not been pursued further. Instead, actors within and outside of the municipality have carried the work forward via commercial sharing projects (e.g. for-profit sharing platforms), civil-society movements (e.g. non-profit USOs) and the initiatives of municipal organisations and sub-units (Bezirk).

London published the Smart London Plan (SLP) in 2013 as a response to its population growth, which the city expects will exacerbate congestion and the resulting air pollution, and increase the strain on healthcare and the management of utilities (Mayor of London, 2013). To deliver solutions inclusive of all London citizens, the plan stresses the importance of collaboration between citizens, businesses, researchers, investors and other stakeholders. However, while the city is making efforts to engage citizens; for example, by managing the Talk London website, which invites citizens to help design policies; the focus of the SLP has been mainly on collaboration with businesses to deliver technological innovations, such as street lighting or smart congestion charges. At the same time, one of the objectives of the SLP is promoting the start-up scene in London by operating a number of support programmes, such as the online platform for tech start-ups Tech.London, or the information and statistics website London DataStore. USOs are included in the GLA's efforts to support innovative start-ups, but the city currently does not have a sharing agenda. The majority of USOs has developed independently of the local government, although mobility organisations have received much attention from the GLA, which is responsible for operating Transport for London (TfL) – the public transport authority in London, and is heavily involved in the smart mobility plan for London.

In this section, empirical data on the governance of USOs in Berlin and London are analysed in line with the conceptual framework developed in section 2. The four distinct roles that a city may undertake in governing USOs are discussed.

City as regulator

Efforts to regulate USOs in both London and Berlin seem to correlate with the size of USOs and the degree of public interest in them. Large for-profit organisations, such as Airbnb and Uber, have received much regulatory attention in Berlin in 2015, and thereafter also in London. The difference is that London has enforced these regulations on a city level, while the enforcement in Berlin remains fragmented and is up to each Bezirk. In some Bezirks, the operations of both platforms are now inhibited by regulations with the motivation that some forms of sharing, including intensive renting of private apartments, have negative consequences for the city and its citizens because they increase gentrification. An interviewee from a B2C car sharing organisation in London, who spends considerable time in discussions with the local government, speaks similarly about the attitude of the city government towards regulating USOs:

> The city is very well aware that some of these innovations will be positive, while others may come with downsides and there is need to make sure that the city understands both (…) They're trying to remain outward facing, looking at the innovations as they come along (…). They evaluate each on its merits and make sure that what they deliver is good for our city.

Both municipalities were found to introduce regulations to reduce the negative impacts of specific USOs, rather than employing one-size-fits-all regulations. In 2017, TfL suspended the licence of the peer-to-peer (P2P) car sharing organisation Uber. TfL deemed the company as "not fit and proper" to hold the licence primarily on the grounds of its approaches to reporting serious criminal offences, obtaining medical certificates, information disclosure and unsatisfactory software standards "able to block regulatory bodies from gaining full access to the app and prevent officials from undertaking regulatory or law enforcement duties" (TfL 2017). Now, all companies operating a P2P taxi service in London must comply with a new regulation, tailor-made for them, while the incumbent taxi industry is regulated by a different, much stricter scheme called the black cab regulation. Interestingly, car sharing clubs, including Carplus and Zipcar, are regulated by licences issued by individual London boroughs. A car sharing club that complies with the rules can be issued a licence to operate.

Food sharing organisations in Berlin and London are also regulated based on their business models. In Berlin, a non-profit organisation that came into focus of regulators was *Foodsharing.de*. One of their food saving initiatives, freely accessible fridges and cupboards in the urban space received a lot of positive media attention. However, the health authorities in a district municipal body Bezirk Pankow demanded a formal structure around the initiative to guarantee food safety. After much discussions and a public conflict between the responsible city representatives ("Lebensmittelaufsicht"), the public food exchanges in the district had to close.

While Foodsharing.de is fighting regulation in Berlin, the London-based food sharing social enterprise Olio complained about the lack of regulatory directions in London. After having contacted the borough of Islington's Food and Safety Authority, Olio was assured that they were not bound by any regulations. At the same time, Olio registered as a food business and applied for a food licence, to make sure it complies with any future regulations. In Berlin, we found another USO that expressed their frustration about the lack of legal clarity. This USO facilitates online sharing of things (anything from video games to drills and cars). According to its founder, there has always been legal ambiguity, but with the rapid development of ICT, it has reached a noticeable impact:

> There is a total legal uncertainty about when the activity starts to be commercial and when it remains private. If I rent out my car on Drivy five times, it is private, but if I rent it out ten times it becomes commercial. This is totally (…) unclear.

Legal ambiguity was also found to be an issue for an interviewee from a P2P accommodation sharing platform, who believes that the regulatory framework in Berlin is strict, as well as vague. Currently, home owners in Berlin can rent out no more than 50% of their apartments without a permit. The law, enforced by a €100,000 fine, only applies to Airbnb-type of rentals, while home exchanges are not affected by it (this also demonstrates the case-by-case approach to regulating USOs).

> It's a very, very strict rule. (…) It makes any short-term accommodation subject to permits. (…) Everybody agrees that [the law] was designed and should be targeted at unwelcome, commercial operators who are misusing private accommodation and just renting it year-round to tourists. (…) But (…) there is a huge divergence in perspectives. (…) Some district officials and the elected politicians in the municipal districts of Berlin would tell you that the law doesn't apply to somebody's primary home. Others would say it does apply to primary homes but you can get a permit for it. Others would tell you that it does apply and you can't get a permit because you're not allowed to do it. And others just have no idea. So it's difficult to reach consensus on what the law does and doesn't cover. (…) There is now total confusion about when you're allowed to host, how often you're allowed to host, whether you need a permit, what the grounds are and which district will grant these permits, and the whole thing is a total mess.

Interviewees from much smaller USOs in Berlin expressed their opinion that their platforms were too small to be of interest for authorities, but believed that their growth would lead to stricter municipal scrutiny. They hoped that they would remain outside of the municipality's focus so their platform could develop further within the current regulatory settings.

It appears that the willingness of local governments to engage in a discussion with USOs about regulations differs in London and in Berlin. Although we found one case where a Berlin city district engaged in a discussion with a USO and changed its attitude from restrictive to supportive, overall, the municipality turns to regulatory modes of government when dealing with USOs, and is not very willing to negotiate with them. This could be attributed to the de-homogenised governmental set-up of Berlin, which contributes to a high level of bureaucracy. The non-profit initiative Mundraub, which runs an online platform mapping publically available edible plants, and which received an award by the German Environmental Protection Agency, described the bureaucratic reality it faces when dealing with municipal sub-units: "Indeed, we have to go from Bezirk to Bezirk to present our project, and surely it will take a while for a decision to be made every time (…). The process looks different in every Bezirk."

In contrast, an interviewee from a car sharing USO in London believed that regulations imposed by the local government in London were not too restrictive: "Uber and Airbnb are banned by the legislation in some countries (…), whereas here, it is more of a two-way conversation, which is far more collaborative and productive." This view was also supported by an interviewee from an accommodation sharing platform, who appreciated that, compared to Berlin, London city officials are open to discussing short-term rentals and work together with USOs. For example, local authorities negotiated with Airbnb to change the formerly very restrictive regulation regarding short-term lettings.[2] Our interviewee appreciates that it gives a clear direction on what can be rented: "The position in London from 1973 onwards was that all short-term rentals needed to get a permission. So, (…) the 90-night rule is an enabling legislation, not a restricting legislation."

The two-way communication between London city governments and USOs is further demonstrated by a discussion between London city councils and an organisation that connects drivers with owners of empty parking spaces:

> We did have a few cases with the councils where [they] were threatening to fine us and proceed with legal action against people renting out their driveways because they claimed it was turning their home into a business. But through negotiations and discussions with councils, we managed to get that legislation reversed.

The aforementioned examples demonstrate the attempts of USOs to engage in collaborative urban governance and change regulations in their favour through negotiations with city councils.

This mode of governance, however, still represents the old tradition of a top-down policy approach when the city authorities make selective decisions in relation to certain USOs. Another way for USOs to engage in the co-production of regulations and policies is through joining efforts with other USOs and delegating third-party actors to negotiate with city governments on their behalf. While examples of such action exist (e.g. a car sharing association in Berlin lobbying for changes in parking regulations, or the industry association Sharing Economy UK based in London representing the voice of USOs), the co-production of regulation by USOs remains nascent.

In summary, in their role of *city as regulator*, the cities tend to focus on selected USOs. Both cities mainly regulate "the big" and "the loud," such as Airbnb and Uber, while neglecting smaller USOs. USOs themselves report struggling with legal ambiguity in both cities, which can be explained by the novelty of the urban sharing phenomenon. While London seems to be open to negotiations with USOs, the Berlin municipality was found to be less interested in dialogue. Efforts of third-party actors to represent the interests of USOs in city governments and to co-produce policies and regulations are emerging, but remain rather marginal.

City as provider

One of the roles a city can take to support USOs is that of an investor. However, financial support for niche start-up initiatives has not been easily available in either of our case cities. Thus, cities have had to find creative ways to support urban sharing.

In the U.K., sharing is promoted through a tax break for users of sharing platforms across the U.K., which was introduced by the national government. According to one of the interviewees, it was based on the needs of London citizens and was pushed for by London's politicians. The overall tax-free allowance is £2000 with a maximum of £1000 per USO. This means that users can, for example, earn £1000 on Airbnb and £1000 on JustPark before having to declare income from these sources.

Some of our interviewees in both cities believe that the main reason for the lack of funding options is large financial cuts for local authorities. This view was supported by Berlin city officials, who stated that the funding they could potentially give to USOs was very small:

> We do not have the manpower or the financial means to be proactive (…). The official position is that we should support this (sharing as waste prevention), but it is not supported with the necessary manpower and funding (…). Only if (supporting a project) does not cost anything and the employee is willing to put in the time, then a fast decision is possible.

Among municipal funding options for urban sharing in Berlin, the most prominent is the funding for business start-ups through the Investment Bank Berlin (IBB), owned by the "Land Berlin".[3] However, our interviewees in Berlin revealed a low level of interest in direct financial support from the city government. Some sharing organisations focused on funding from other sources, such as the federal government, crowdfunding or the EU. Overall, the interviews in Berlin revealed a low level of direct financial support for sharing organisations.

While London also has limited financial means, an interviewee from the GLA said that the city supports "(…) new innovations in the sharing and collaborative economy that tackle London's challenges in many ways, from crowdfunding, innovation investment, and co-creation of digital products with Londoners [to] increase their access to public services and information."

However, similarly to Berlin, the city government is not running any funding programme directly aimed at USOs. Instead, much of the funding is open to all start-ups, including USOs. Of interest are two Mayor's projects supporting innovation: the London Co-investment Fund and the Mayor's Crowdfunding Pilot. The former is financed jointly by the London Mayor and the London's Enterprise Panel. It has sponsored a number of tech start-ups, including two USOs: Hubble, a business-to-business online platform for office space and Flat Club, an online platform for medium-term stays. The Crowdfunding Pilot offers an online platform where local groups can propose and crowdfund community project ideas. Once a project reaches a desired number of backers, the Mayor pledges the remaining money needed to develop it. According to an interviewee from the GLA, the Crowdfunding Pilot "(…) was recently singled out by the World Government Summit as one of the leading government innovations from around the world." While the project has potential to fund future sharing projects, no USO has received funding yet.

An organisation directly targeting USOs in the U.K. is Nesta, a London-based independent charity supporting innovation. It receives funding from corporations, foreign governments, local governments and charities, and has recently launched the ShareLabFund with the aim to diversify the portfolio of collaborative economy platforms in the U.K., and to support the development of sharing innovations based on the ideas of collaborative consumption, production and learning. It is offering between £10,000 and £40,000 to sharing organisations that deliver public services and social impact.

Apart from offering funding, municipalities can also support USOs through the provision of premises or space. For example, London boroughs are granting parking spaces to car clubs. Compared to the general fleet, car clubs have a more environmentally sound fleet (Bundesverband Carsharing 2010), thus contributing to the aim of GLA to tackle air pollution in the city. An interviewee from the GLA noted:

> London is now getting completely away from private car ownership. The idea of renting cars is taking hold (…). Londoners are happy to share a car with each other because we are happy to sit on the public transport with each other as well.

However, boroughs are selective in allowing access to parking spaces. For example, the City of Westminster[4] has dedicated parking spaces to a sole car club operator, Zipcar. At the same time, Zipcar has been collaborating with the city government and TfL on the next public transport strategy. It has also been lobbying for the introduction of congestion charges for all private vehicles as well as for taxation of private cars in London. The company uses data about car sharing collected by TfL and the car sharing sector to make their benefit for the municipality evident:

> We wanted to prove that the outcomes of car sharing were positive for the city. The studies show that after joining a car club, people drive less compared to when they owned their own car and they take sustainable modes of transport: walking, cycling, public transport, more. So we managed to prove to them that car sharing has merits for the city. The studies were used by the TfL to fund the London boroughs to put in the parking spaces. London takes a data-driven approach to what is good for it so it was very much in the sector's best interest to demonstrate with data how positive our service was.

However, not all types of car sharing clubs are treated equally in London. Currently, there are two business-to-consumer car sharing models operating in London, a station-based model[5] (Carplus and Zipcar), and an A-to-B model[6] (DriveNow). The station-based model is encouraged by London councils, and was given parking permissions across all of London boroughs. On the other hand, the A-to-B model has only succeeded in North-East London as it was not successful in negotiating parking permissions with the councils in other parts of the city. The difficulty of expanding the A-to-B car sharing model is attributed to insufficient communication between London boroughs, and absence of a London-wide strategy for the allocation of parking spaces.

In Berlin, the role of city as provider can be illustrated by its involvement with issues about premises and space. Our interviewees discussed the problem of increasing gentrification that has led to growing property prices in Berlin, making it challenging to develop creative ideas and continue their operations. The city of Berlin was mentioned as often favouring commercial interests over civil-society interests in conflicts over property access. However, where enough local pressure was generated by well-organised civil-society and inhabitants, Berlin municipality had been cooperative and ensured that property or space remained available to civil-society organisations.

In summary, the role of *city as provider* can be divided into two sub-sections: city as investor and city as host. The role of the city as investor is currently limited in both cities. In Berlin, USOs receive financial support from federal government and EU funds or through crowdfunding. In London, there are two funding projects run by the Mayor, as well as opportunities to receive finance through crowdfunding and charities, such as Nesta, which supports innovative ideas in the U.K. In its role as a host, the city of London provides parking spaces to car sharing companies. Some London boroughs remain selective in terms of granting access to public parking spaces, which makes the expansion of an A-to-B car sharing model somewhat contested. In Berlin, when civil-society groups and citizens lobbied for access to property rights, the city government demonstrated higher degrees of co-operation and inclusiveness.

City as enabler

According to the interviewees, there are hopes from USOs that municipalities would take a more proactive role as enablers for their operations. The sharing activists in Berlin own the domain www.sharingcityberlin.org, but they would like the city to take charge of it and develop a broader political vision for urban sharing in Berlin. The interviewees expressed a wish for the municipality to organise workshops and other events for the sharing movement in Berlin. They called for more clarity about who to contact in the city government about issues regarding urban sharing. Only one for-profit USO interviewed in this study was satisfied with the amount of support from the municipality, but conceded that the support came from the business development division of the municipality, which focuses on business start-ups. Interestingly, the interviewees felt that the Berlin municipality showed a higher interest in the sharing idea and development of USOs before 2015 when the municipality funded a feasibility study to investigate the potential of the sharing economy in Berlin, and organised workshops with stakeholders. However, the study was never published, and municipal support of sharing disappeared soon thereafter.

Current municipal support of USOs in Berlin appears mainly at a district level. The previously mentioned "Bezirk Pankow" is a good example of city engagement with food sharing. Officials working for this district labelled it as an "edible area" with a public profile and a homepage and funded a non-profit sharing organisation to run the homepage. Furthermore, the organisation was able to access data about edible plants from the municipality and integrate it into their homepage.

Due to the perceived lack of municipal willingness to coordinate USOs in Berlin, civil-society organisations, primarily OuiShare, were described as the coordinating and driving force behind the development of urban sharing in Berlin. This, however, was primarily true for non-commercial USOs, while for-profit organisations perceived that they received better support from the city. In this sense, there seems to be a reinforced justice dilemma with Berlin municipality being more supportive of commercial USOs rather than non-profit ones, which could have negative implications for social sustainability in the city.

All of the interviewed USOs facilitate sharing through an online platform. Consequently, creating online content was identified as being of critical importance by several of the USOs based in Berlin. Many of them sought assistance in terms of expertise and funding to make their online appearance more professional:

> I plan to expand (the platform) so that free things can be identified and additional services offered (…) I have a whole set of add-ons, and I am talking to a number of (programmers) from Common Bookings Tool to implement it.

Perhaps a lesson about fostering ICT knowledge can be learnt from the city of London. As part of the smart city agenda, London organises hackathons for entrepreneurs to learn from each other and share their ICT skills. As a smart city, London is also creating extensive data infrastructure in a project called Sharing Cities. The GLA envisions the data infrastructure to be useful for many of the London start-ups, including those specialising in sharing. The data will be accessible to them for free or for a small fee.

Similarly to Berlin, there are many expectations from London-based USOs about the ways London can enable their activities. For example, sharing clubs have benefited from GLA's and TfL's overall willingness to support car sharing as the GLA gives them strategic direction and helps promote their

activity. On the other hand, other USOs find it difficult to catch the attention of municipal governments. The previously mentioned food sharing organisation Olio seeks to become more visible on
local authorities' websites, but has only been successful outside of London; and a time sharing organisation Echo has helped a local council team, which supports innovation in the area, build a P2P
business network, but has not received incentives from the council in return. A London interviewee
from the municipality believes that the lack of interest from the councils is one of the reasons behind
the slow development of small, non-profit sharing platforms: "I've seen people trying to set up tool-
sharing initiatives but when they try to find people to help and support them, councils do not feel like
it is their responsibility."

In Berlin, recognition by the municipality was among the most favoured ways of municipal support
expressed by the interviewees. However, several interviewees voiced their frustration about the
municipality's lack of knowledge about USOs' efforts. This was particularly important as online
sharing often takes place among strangers and the platform depends on trustworthiness. The
USOs therefore hoped that if the municipality recognised them, they would reach more participants
in the city:

> One needs marketing; one needs money. We are a small business and we only got a small amount of investment.
> That money is gone and we will need to see how to develop further. Sharing is very marketing intense. One needs
> to reach to private people, and this costs money, and therefore we are not able to implement the [P2P] concept as
> we thought.

In London, the platforms JustPark and GetTuxi tell a different story. Thanks to the SLP, they have
been able to benefit from direct attention from the Mayor of London:

> We were representing London at the Smart City Expo World Congress (…) in Barcelona, championing tech and
> smart cities (…) [B]eing selected by the Mayor of London was a big step for us. There was a London Smart City
> stand together with GetTuxi and a few other forward-thinking tech start-ups, which represent the Mayor of
> London's prerogative of London as a smart city.

While these two larger, established USOs received support from London government, their
smaller, not-for-profit counterparts still struggle to get attention from the municipality. Contrary to
both London and Berlin promoting themselves as smart cities, neither of the city governments
describes itself as a sharing city and consequently does not have dedicated employees working
with USOs. An interviewee from an accommodation sharing platform stated:

> A couple of years ago, there was lots of talk about sharing in the cities and shareable cities, and that's still going
> on, but that thought got mashed together with smart cities as if it was the same thing, and it's obviously not.
> We've never had a discussion in London, nor in Berlin, where it's made clear what their specific definition is,
> or what part we can play in it, if any.

Similarly, another interviewee[7] from a civil-society group did not perceive the concept of smart
cities as facilitating any real development of USOs:

> The majority of smart city agenda in London is focused on data collection and analysis, and not on collaborative
> economy and sharing. That is why London is not at the forefront of smart cities. It is a very traditional smart city
> project. It is about a better flow of traffic and people to lower the negative impact on the city and the
> environment.

In summary, the role of *city as enabler* can be divided into two roles: city as matchmaker and city as
communicator. In Berlin, USOs expressed the need for a more proactive city government, and a more
active role of the city as matchmaker and organiser of workshops and sharing events. Berlin was more
active in the realm of sharing before 2015, but current municipal support appears to take place
mainly at a district level. In London, larger for-profit USOs were more likely to gain municipal endorsement, while non-profit USOs went unnoticed. In its role as a matchmaker and under the smart city
agenda, London organises hackathons and runs the Sharing Cities project. Both Berlin and London-
based USOs seem to expect a much more active city role as an enabler. Another important

observation is that the fragmented governmental structure of both cities makes city-wide support for any given USO a challenging task.

City as consumer

The lack of a coherent and committed strategy towards sharing in the city of Berlin is probably a key reason behind the absence of any attempts by the municipality to engage with USOs for internal purposes. Neither does the municipality exercise its role as a role model in visualising the feasibility of sharing innovations.

When analysing the role of Berlin as a consumer, it is worth highlighting one peculiar case of the *Berliner Stadtreinigung* (BSR), a quasi-public institution under the regional government of Berlin, which is in charge of waste handling in the city. The BSR's work is informed by the regulatory guidance from the EU, as well as by federal and municipal legislation. Legislative focus on waste prevention has led to an online second-hand exchange, as well as the support of some offline-based USOs in Berlin in 2014. BSR also supports a project run by an apartment block association and the organisation *Pumpi Pumpe*, which aims to help neighbours share their resources by distributing door stickers to indicate what things they own and are willing to share. BSR also publishes a magazine where sharing has been discussed several times. While BSR is not a unit of the Berlin municipality in a narrow sense, it is tasked with implementing political decisions and has, in practice, facilitated sharing in Berlin on multiple levels.

The role of London as a consumer of USOs is also rather small, although some councils procure mobility services from commercial car sharing clubs. For example, the Croydon council is collaborating with Zipcar to procure car club services to their employees. The aim of this partnership is to achieve reductions in the council's car usage, CO_2 emissions and travel costs.

In summary, neither London nor Berlin have any pilot or demonstration project on sharing that would demonstrate the city's *role as consumer*, although they do experiment with smart city pilots.

Summary of findings

Our conceptual framework draws on the literature on urban climate governance (Bulkeley and Kern 2006; Kern and Alber 2008). It specifies four roles cities may undertake in their efforts to govern ICT-enabled urban sharing: city as regulator, city as provider, city as enabler and city as consumer. The framework proved to be useful for both data collection and analysis, and we were able to confirm the roles cities may assume when governing urban sharing.

Overall, the general conclusion is that both cities indirectly support ICT-enabled urban sharing as part of their smart agenda that aims to facilitate ICT-enabled technical innovation and the emergence of start-ups. Larger, for-profit USOs may receive support, endorsement and sometimes even promotion by the city governments; however, non-profit USOs are often left without support.

At the same time, we found that it is usually larger USOs that come into the focus of city regulators, particularly when they are perceived to be negatively affecting the cities and their citizens. This is sometimes also the case for small USOs, especially when they attract media attention and for example public health and safety is brought to the attention of the municipality.

Overall, our comparative analysis shows that London and Berlin assume different roles in their engagement with the urban sharing phenomenon. We found examples that help operationalise the general roles and modes of municipal governance. The conceptual framework can therefore be customised to urban sharing.

Discussion and conclusions

A shared perception of our interviewees from USOs, municipalities and third-party organisations was that the cities do not relate to "urban sharing" or "the sharing economy" as such. Instead, they engage with USOs as with other business start-ups. However, if the USOs are part of waste prevention initiatives (Berlin) or mobility-related solutions (London), they tend to attract more attention and support from the local governments. The crucial aspect that motivates Berlin municipality to offer support for USOs is the potential impact in terms of innovation and economic development. This makes it very difficult to discuss the possible contributions of the sharing concept to the city. The drawback is somewhat overcome by the vibrant landscape of civil-society organisations working with urban sharing, partly by soliciting support from a higher political (state) level, and party by looking towards other municipalities, which are more positive towards sharing. London, on the other hand, seems to take more of a pragmatic and collaborative approach with USOs, and is open to, for example, discuss issues around regulation. The city representatives understand that some USOs will have positive impact, while others might give raise to problems. Thus, they support or regulate USOs based on a case-by-case basis.

Our study shows that some modes of urban sharing governance are more prominent than others. For example, both cities often engage with urban sharing in their role as regulators, following their mandate to represent and guard the interests of the city and its inhabitants regarding, for instance, quality of life, economic prosperity, environmental quality, social justice or health and safety.

This happens, however, in a traditional manner of top-down policy approach, while the role of third-party actors lobbying for the interests of USOs in city governments remains marginal. At the same time, unionising into inter-field (e.g. mobility USOs) and intra-field (e.g. mobility with accommodation USOs) networks could ensure a transition to more efficient and inclusive collaborative urban governance processes. Such processes would empower USOs by making their voices heard by the city councils, which in turn would support the institutionalisation of the urban sharing phenomenon.

The role of city as provider is currently rather limited, especially in its role as an investor. Both cities undertake a slightly more prominent role as hosts by providing premises or spaces to selected USOs. So far, we have identified only a handful of strategic projects that promote urban sharing in London, while most of the supporting and enabling activities in Berlin take place at the local, district level. We only found one example of city as consumer and a role model, which means that there is clearly an opportunity for improvement if cities are interested to capitalise on urban sharing within their smart agendas, or to advance waste reduction, social cohesion, improve air quality or fulfil other sustainability ambitions.

Both cities appeared to be interested in USOs due to their environmental and social potentials. While sharing is not officially talked about in the municipalities of Berlin and London as a tool for sustainable development, it is acknowledged as such in various cases of concrete action to promote sustainability. Perhaps, the most prominent example is car sharing. Despite the fact that both cities took decisive action to regulate Uber, they nonetheless acknowledge the potential of other types of car sharing for sustainable urban development. In early 2017, the new coalition government in Berlin consisting of the Social Democratic party, the Left party and the Greens expressed challenge for the municipality to distinguish car sharing that is beneficial for the municipality (i.e. reduced traffic or parking space requirements) and that is potentially harmful (such as A-to-B car sharing). In London, car sharing organisations such as CarPlus, Zipcar, or DriveNow thrive and enjoy an overwhelming support from the GLA. It became apparent from the interviews that the municipality supports car sharing in London for its potential to decrease congestion and air pollution.

Another way both cities are dealing with air pollution is through their bike sharing schemes. Berlin is an increasingly bike-friendly city, and acknowledges bike sharing as an essential part of this development. "Bike sharing is quite strongly supported by the municipality, both through various public services, but also the support of private services. Sure, there is a lot more potential, but the municipality looks at it in a favourable way." (Municipal employee)

The Berlin municipality also supports other types of sharing projects, such as food sharing and stuff sharing, when it believes it can improve urban sustainability.

As discussed earlier, if managed properly, USOs have a potential to contribute to sustainable urban development, and advance social justice and environmental sustainability. Although the goal of both cities is to support sustainable innovation as part of their smart city agendas, the support for USOs, which could potentially lead to improved environmental and social conditions in the cities, remains fragmented. Often, USOs are lumped together with smart, innovative start-ups, which could possibly lead to their sustainability potentials remaining untapped.

Both cities have succeeded in regulating unwanted, disruptive activities on a case-by-case basis. However, the regulatory action was not balanced with systems of provisions or enabling policies targeted at USOs, which hold the potential to bring environmental and social justice to cities. Therefore, we suggest cities continue regulating larger USOs on a case-by-case basis, but also introduce supportive mechanisms for smaller, bottom-up USOs, which hold social justice at the core of their operations. This intervention would even up the level-playing field between small USOs and large-scale USOs. A failure to distinguish among sizes and purposes of USOs risks that cities will become the dystopian future that Hollands (2015, 73) describes when he argues that

> [w]e should be wary of corporately inspired smart scenarios where urban problems have all been solved by technology and all of its inhabitants are happy and prosperious; however, tantalising this vision is. Underlying this idea is a more manipulative notion that cities are just "machines for making money out of" or that global competitiveness between cities will automatically make them better places to live.

Urban sharing is interpreted in many ways in London and Berlin, but if the focus of municipalities remains on larger platforms, there is a risk of streamlining sharing to mean primarily "making money"; a development that suppresses many other functions USOs can fulfil in an urban space.

As many USOs complained about unclear rules, efforts could also be made by the municipalities to give a clear signal about which types of USOs are welcome in the city, and which are not. Municipalities also have the potential to expand their role as consumers and become role models by directly engaging with sharing, but we found that they did not take advantage of this mode of governance. As an inspiration, this suggestion can be demonstrated on a case from Malmö, Sweden, where the municipality developed an online platform where public organisations (e.g. such as schools or hospitals) can share unwanted and surplus furniture and office equipment. The platform allows them to reduce economic and environmental costs by utilising the idling capacity of second-hand assets.

Interestingly, when we compare the prominence of certain roles in our urban sharing cases with urban climate governance literature, we see significant differences. For example, Kern and Alber (2009, 171) report that "municipalities do not fully exploit their authoritative powers and are reluctant to apply authoritative modes of governing through regulatory measures and strategic planning". On the other hand, our findings suggest that both cities are actively engaged in developing regulations for USOs.

Furthermore, in climate governance, municipalities undertake prominent roles as facilitators, but we our findings show that the role of a municipality as a facilitator of sharing is quite limited. The reasons for these divergent situations are worth exploring in future research.

Overall, we found that the smart city agendas are not strategically coupled with urban sharing as cities do not have explicit policies or strategies for urban sharing. Although their smart city agendas might mention urban sharing as an example of how businesses and grassroots sharing organisations capitalise on the penetration of ICT infrastructure and on the availability of ICT solutions, they lack strategic engagement with USOs.

We can conclude that cities indirectly support USOs as innovations that can contribute to smart city agendas, and inhibit them when they are perceived as having a negative impact on their citizens. Future research could potentially explore the justice dilemma that emerges in the light of cities supporting larger USOs, which promise to advance economic development, while neglecting smaller USOs, which often adhere to the principles of environmental sustainability and social justice. Another direction for future research is to explore how the roles of cities in sharing and smart agendas are changing over time, identifying dynamic institutional processes and reasons behind them.

Notes

1. In this paper, we use the term 'urban sharing' instead of 'the sharing economy' to discuss our contribution as we see it as less delimiting in terms of the types of sharing initiatives included under its umbrella.
2. The regulation allows/restricts "entire home" lettings to 90 days in a calendar year. It does not apply to rooms in a property where people normally live.
3. Berlin is both 'Land' and 'Gemeinde' at the same time, combining two legislative levels in Germany.
4. The City of Westminster is one of London's boroughs, but holds a city status.
5. In a station-based model, a car is picked up and delivered at the same place.
6. In an A-to-B model, a car can be picked up at the nearest pick-up point and left in another part of the city.
7. This interviewee departs from the notion that smart cities are not merely ICT connected hubs but should also be built on collaboration and sharing between their citizens.

Disclosure statement

No potential conflict of interest was reported by the authors.

Funding

This paper is a result of two projects at the IIIEE at Lund University: Urban Sharing funded by Riksbankens Jubileumsfond and Urban Reconomy funded by the Swedish Research Council FORMAS.

ORCID

Lucie Zvolska http://orcid.org/0000-0003-1660-1452

References

Agyeman, J., D. McLaren, and A. Schaefer-Borrego. 2013. *Sharing Cities*. https://friendsoftheearth.uk/sites/default/files/downloads/agyeman_sharing_cities.pdf.
BBC. 2016. "Airbnb announce plans to block hosts exceeding 90-day rental limit." *BBC*, December 1. http://www.bbc.com/news/uk-england-london-38169788.
Bulkeley, H., and V. Castán Broto. 2013. "Government by Experiment? Global Cities and the Governing of Climate Change." *Transactions of the Institute of British Geographers* 38 (3): 361–375.
Bulkeley, H., V. Castán Broto, M. Hodson, and S. Marvin. 2010. *Cities and Low Carbon Transitions*. New York: Routledge.
Bulkeley, H., and K. Kern. 2006. "Local Government and the Governing of Climate Change in Germany and the UK." *Urban Studies* 43 (12): 2237–2259. doi:10.1080/00420980600936491.
Bundesverband CarSharing. 2010. *The State of European Car-Sharing. Final Report D 2.4 Work Package 2*. https://www.motiva.fi/files/4138/WP2_Final_Report.pdf.
Cattacin, S., and A. Zimmer. 2016. "Urban Governance and Social Innovations." In *Social Innovations in the Urban Context*, edited by T. Brandsen, S. Cattacin, A. Evers, and A. Zimmer, 21–44. Heidelberg: Springer.
Cohen, M. J. 2016. *Sharing in the New Economy the Future of Consumer Society: Prospects for Sustainability in the New Economy*. Oxford: Oxford University Press.
Cohen, B., and J. Kietzmann. 2014. "Ride on! Mobility Business Models for the Sharing Economy." *Organization & Environment* 27 (3): 279–296.
Dyer, M., D. Gleeson, and T. Grey. 2017. "Framework for Collaborative Urbanism." In *Citizen Empowerment and Innovation in the Data-Rich City*, edited by C. Certomà, M. Dyer, L. Pocatilu, and F. Rizzi, 19–30. Tracts in Civil Engineering. Cham: Springer.
EC. 2013. *Smart Cities Stakeholder Platform: Integrated Action Plan – Report Process & Guidelines*. http://archive.eu-smartcities.eu/sites/all/files/Integrated20Action20Plan.pdf.
Evans, J., and A. Karvonen. 2010. "Living Laboratories for Sustainability: Exploring the Politics and Epistemology of Urban Transition." In *Cities and Low Carbon Transitions*, edited by H. Bulkeley, V. Castán Broto, M. Hodson, and S. Marvin, 256. London: Routledge.
Evans, J., and A. Karvonen. 2014. "'Give Me a Laboratory and I Will Lower Your Carbon Footprint!' — Urban Laboratories and the Governance of Low-Carbon Futures." *International Journal of Urban and Regional Research* 38 (2): 413–430. doi:10.1111/1468-2427.12077.
Gibson, J., M. Robinson, and S. Cain. 2015. *CITIE: A Resource for City Leadership*. http://citie.org/reports/.
GLA. 2016. *About Us*. https://www.london.gov.uk/about-us.

Gleeson, D., and M. Dyer. 2017. "Manifesto for Collaborative Urbanism." In *Citizen Empowerment and Innovation in the Data-Rich City* edited by C. Certomà, M. Dyer, L. Pocatilu, and F. Rizzi, 3–18. Tracts in Civil Engineering. Cham: Springer.

Gori, P., P. L. Parcu, and M. L. Stasi. 2015. "Smart Cities and Sharing Economy." Vol. 96, Robert Schuman Centre for Advanced Studies Research. http://cadmus.eui.eu/bitstream/handle/1814/38264/RSCAS_2015_96.pdf?sequence=1.

Grübler, A., and D. Fisk. 2013. *Energizing Sustainable Cities: Assessing Urban Energy.* Abingdon: Routledge.

Guttentag, D. 2015. "Airbnb: Disruptive Innovation and the Rise of an Informal Tourism Accommodation Sector." *Current Issues in Tourism* 18 (12): 1192–1217. doi:10.1080/13683500.2013.827159.

Hodson, M., and S. Marvin. 2010. "Can Cities Shape Socio-Technical Transitions and How Would We Know If They Were?" *Research Policy* 39: 477–485.

Hollands, R. G. 2008. "Will the Real Smart City Please Stand up? Intelligent, Progressive or Entrepreneurial?" *City* 12 (3): 303–320. doi:10.1080/13604810802479126.

Hollands, R. G. 2015. "Critical Interventions into the Corporate Smart City." *Cambridge Journal of Regions, Economy and Society* 8 (1): 61–77.

Höjer, M., and Wangel, J. (2015). "Smart Sustainable Cities: Definition and Challenges." In *ICT Innovations for Sustainability. Advances in Intelligent Systems and Computing 310*, edited by L. M. Hilty and B. Aebischer. Springer.

Katz, V. 2015. "Regulating the Sharing Economy." *Berkeley Technology Law Journal* 30: 1067–1126.

Kern, K., and G. Alber. 2008. "Governing Climate Change in Cities: Modes of Urban Climate Governance in Multi-Level Systems." Paper presented at the Competitive Cities and Climate Change.

Kern, K., and G. Alber. 2009. "Governing Climate Change in Cities: Modes of Urban climate governance in multi-level systems." Paper presented at the International Conference on Competitive Cities and Climate Change, Milan, October 9–10.

Khan, J. 2013. "What Role for Network Governance in Urban Low Carbon Transitions?" *Journal of Cleaner Production* 50: 133–139. doi:10.1016/j.jclepro.2012.11.045.

Kronsell, A., and D. Mukhtar Landgren. 2017. Collaborative urban governance through experimentation: Municipalities in Sustainability Transitions through Urban Living Labs (manuscript), 20.

König, A., and J. Evans. 2013. "Experimenting for Sustainable Development? Living Laboratories, Social Learning, and the Role of the University." In *Regenerative Sustainable Development of Universities and Cities: the Role of Living Laboratories*, edited by A. König, 1–24. Cheltenham: Edward Elgar.

Laurell, C., and C. Sandström. 2016. "Analysing Uber in Social Media – Disruptive Technology or Institutional Disruption?" *International Journal of Innovation Management* 20 (5), doi:10.1142/S1363919616400132.

McLaren, D., and J. Agyeman. 2015. *Sharing Cities: A Case for Truly Smart and Sustainable Cities.* Cambridge: MIT Press.

Menny, M., Y. Voytenko Palgan, and K. Mccormick. 2018. "Urban Living Labs and the Role of Users in Co-Creation." *Gaia* 26 (1): 68–77.

OECD. 2008. "Competitive Cities and Climate Change." Paper presented at the Competitive Cities and Climate Change.

Orsi, J. 2013. *"The Sharing Economy Just Got Real."* http://www.shareable.net/blog/the-sharing-economy-just-got-real.

Powell, W. W. 1990. "Neither Market nor Hierarchy: Network Forms of Organization." *Research in Organizational Behaviour* 12: 295–336.

Schor, J. 2014. *Debating the Sharing Economy: Great Transition Initiative.* http://greattransition.org/publication/debating-the-sharing-economy.

Sullivan, C. 2016. "Sadiq Khan Launches Plan to Help London's Black Cabs Tackle Uber." *Financial Times*, September 13. https://www.ft.com.

Sundararajan, A. 2016. *The Sharing Economy: The End of Employment and the Rise of Crowd-Based Capitalism.* Cambridge, MA: MIT Press.

TfL. 2017. *"Licensing Decision on Uber London Limited."* https://tfl.gov.uk/info-for/media/press-releases/2017/september/licensing-decision-on-uber-london-limited.

Trencher, G., X. Bai, J. Evans, K. McCormick, and M. Yarime. 2014. "University Partnerships for co-Designing and Co-producing Urban Sustainability." *Global Environmental Change* 28: 153–165. doi:10.1016/j.gloenvcha.2014.06.009.

UN. 2015. *World Urbanization Prospects: The 2014 Revision.* https://esa.un.org/unpd/wup/Publications/Files/WUP2014-Report.pdf.

UNEP. 2013. *City-Level Decoupling: Urban Resource Flows and the Governance of Infrastructure Transitions.* http://www.unep.org/resourcepanel/portals/24102/pdfs/cities-full_report.pdf.

Vanolo, A. 2014. "Smartmentality: The Smart City as Disciplinary Strategy." *Urban Studies* 51 (5): 883–898.

Wosskow, D. 2014. *Unlocking the Sharing Economy: An Independent Review.* Crown, Department for Business, Innovation and Skills. http://collaborativeeconomy.com/wp/wp-content/uploads/2015/04/Wosskow-D.2014.Unlocking-the-UK-Sharing-Economy.pdf.

Voytenko, Y., K. McCormick, J. Evans, and G. Schliwa. 2016. "Urban Living Labs for Sustainability and low Carbon Cities in Europe: Towards a Research Agenda." *Journal of Cleaner Production* 123: 45–54. doi:10.1016/j.jclepro.2015.08.053.

Appendices

Appendix 1: Interview guides

Sharing organisations
How do you interact with the municipal government?
How does the municipal government encourage your work?
How does the municipal government discourage your work?
How does the local government's position on urban sharing look like?
What are the driving forces for development of sharing initiatives in the city?
How does the local government distinguish between sharing initiatives to encourage and to discourage?
Which are the main constraints for the development of sharing initiatives in the city?
Is there anything else you would like to mention that we have not touched upon yet?
City government
How does the local government's position on urban sharing look like?
How is the work with sharing formalised in the municipality?
What are examples of sharing initiatives that the local government engages in?
How does the local government distinguish between sharing initiatives to encourage and to discourage?
How does the government engage with sharing initiatives (both encourage/discourage)?
What are the driving forces for the development of sharing initiatives in the city?
Which are the main constraints for the development of sharing initiatives in the city?
Is there anything else you would like to mention that we have not touched upon yet?

Appendix 2: Anonymised list of interviewees

Organisation	Position	Location
P2P time sharing organisation	Founder	London
P2P space sharing organisation	Director	London
P2P space[a] sharing organisation	PR & marketing manager	London
P2P food sharing organisation	Manager	London
Municipality	Politician	London
Research Program	Researcher	London
Consultancy	Researcher	London
B2C car sharing organisation	Manager	London
Municipality	Public official	London
B2C space sharing organisation	Founder	London
P2P sharing organisation[b]	Founder	Berlin
P2P sharing organisation	Municipal employee	Berlin
P2P sharing organisation	Founder	Berlin
P2P space sharing organisation	Founder	Berlin
P2P sharing organisation	Founder	Berlin
P2P sharing organisation	Founder	Berlin
P2P food sharing organisation	Leading Member	Berlin
Municipality	Public official working with the sharing city	Berlin
P2P food sharing organisation	Leading member	Berlin
Online news site	Employee who composed a summary of sharing initiatives in Berlin	Berlin
International grassroots group	Leading member	Berlin
P2P bike sharing organisation	Founder	Berlin
Council for Sustainable Development	Employee	Berlin
Municipality	Public official working with sharing	Berlin

[a]"Space" refers to rooms, apartments/houses, offices, or parking spaces.
[b]"Sharing organisation" refers to any smaller items that can be shared, from kitchen, garden and sports equipment to books and toys.

Smart and eco-cities in India and China

Johanna I. Höffken and Agnes Limmer

ABSTRACT

Smart and eco-cities have become important notions for thinking about urban futures. This article contributes to these ongoing debates about smart and eco-urbanism by focussing on recent urbanisation initiatives in Asia. Our study of India's Smart Cities Mission launched under the administration of Narendra Modi and China's All-In-One eco-cities project initiated by Xi Jinpin unfolds in two corresponding narratives. Roy and Ong's [2011. *Worlding Cities: Asian Experiments and the Art of Being Global.* Oxford: Wiley-Blackwell] "worlding cities" serves as the theoretical backdrop of our analysis. Based on a careful review of a diverse set of academic literature, policy and other sources we identify five process-dimensions for analysing the respective urban approaches. We show how the specific features of China's and India's urban focus, organisation, implementation, governance and embedding manifest both nations' approaches to smart and eco-urbanism. We argue that India's Smart City Mission and China's All-in-One project are firmly anchored in broader agendas of change that are set out to transform the nation and extend into time. The Indian Smart City Mission is part of a broader ambition to transform the nation enabling her "smart incarnation" in modernity. Smart technologies are seen as the key drivers of change. In China the framework of ecological civilisation continues a 5000-year historical tradition of civilisation excellence. By explicitly linking eco-urbanism to the framework, eco-cities become a means to enact ecological civilisation on the (urban) ground.

1. Introduction

Cities have always been the sites of aspirations and becoming. The promise of cities is especially luring in the East, where China and India enact their urban ambitions in the context of a drive to shape their identity and future within a global (market) arena. Together, China and India have a population of 2.75 billion people and their populations are expected to continue growing. Currently, 40% of the population in the broader Asian region are below the age of 24 (United Nations – Department of Economic and Social Affairs 2017) and many of these people are striving to live in urban settings, so that projections assume that by 2030 the number of Asian cities with half a million people will increase by 30% (United Nations – Department of Economic and Social Affairs 2016). These dynamics put China's and India's infrastructure, environment and societies under stress. Within these dynamics of a growing population, significant urbanisation trends, economic growth aspirations and the accompanying sustainability challenges, India and China are trying to find answers that will guide them on their way through the twenty-first century. In this article we will focus on India and China's recent urbanisation efforts as this offers an entry into exploring their (urban) visions and answers that seem to emerge from them.

Although an established paradigm for both China and India, "urbanisation" quickly headed their political agendas when Xi Jinping and Narendra Modi took office. Equally rapid, both heads of state announced new urban, smart and green initiatives under whose umbrella the individual nation should be transformed and elevated to a new level of civilisation. However, such rapid development raises challenges for social and environmental equity that need addressing, both carefully and complementary (Agyeman, Bullard, and Evans 2003). Especially, as India's "Smart Cities Mission" and China's "All-In-One Project" on eco-city development are not only striving to work on a national level, but are embedded in global economic and geopolitical networks (Roy and Ong 2011). And they eventually respond to questions of global development and climate change, while clearly aiming to place the two countries on the map of a historical "first world".

China's and India's urban initiatives are, thus, embedded in broader debates on global urbanism, which aims to theorise urban–global relations (e.g. Ong and Collier 2005; Brenner and Keil 2006; McFarlane 2010; Roy and Ong 2011; McCann, Roy, and Ward 2013). They also relate to debates on eco-urbanism (e.g. De Jong et al. 2016; Sharifi 2016) in which cities are often framed as sites where consequences of climate change coalesce (Caprotti 2014). Besides the urban–ecological nexus, another stream of scholarship informing this analysis is the smart urbanism debate, in which cities are analysed with a special focus on new technologies and infrastructures (Hollands 2008; Söderström, Paasche, and Klauser 2014; Vanolo 2014; Marvin, Luque-Ayala, and McFarlane 2016).

In this article we will analyse the initiative's of India's Smart Cities Mission and China's All-In-One eco-cities project against the theoretical backdrop of "worlding cities", developed by Roy and Ong (2011). Worlding directs attention "to identify the projects and practices that instantiate some vision of the world in formation" (Ong 2011, 11). By detailing their respective worlding projects we show how "the urban experience" in both India and China becomes "the ongoing result and target of [the countries'] specific worldings" (Ong 2011, 12).

Focussing on China's and India's urban approaches engages them in a comparative relationship. McFarlane (2010) argues that "urbanism has always been conceived comparatively" (725), not least due to the fact that statements about cities are most often derived from implicit comparisons with other cities. Urban studies scholars have long been engaged in methodological-theoretical discussions about the issue of comparison (e.g. Pickvance 1986; Brenner 2001; Robinson 2011). One of the aspects of these diverse and stimulating reflections highlights how cities of the global North have long been unreflectively taken as the default or benchmark for comparisons. As a result, theory production is severely limited "as variables or topics to be considered [are restricted] to those relevant to the privileged locations" (Robinson 2011, 10). Our focus on Chinese and Indian urban efforts contributes to the body of work that extends the range of cases to other geographies.

Another, yet related, topic of critical concern has been the restricted choice of comparative markers, which are often based on generic economic, political and/or territorial categories. Robinson (2011) suggests to use categories of comparison that are not assumed a priori, but carefully developed in relation to the specifics of the research case(s). Many phenomena cannot be captured by the often reductionist use of economic, political and territorial divides. In these cases, she pleads for units of analysis that operate on a different scale or go "beyond the city's physical or territorial extent" (14). Our work is informed by these theoretical–methodological considerations.

In the analysis of India's Smart Cities Mission and China's All-in-One eco-cities project core policy documents, websites and reports from public and private organisations as well as news and media articles have been carefully selected and their content thoroughly analysed. Besides this, we could complement and contextualise the analysis by drawing on our respective extended and ongoing qualitative research work (including field observations and in/formal stakeholder conversations) in both countries. Through inductive coding of these data sources we identified themes which we interpreted using scholarly literature, mainly within the fields of critical urban studies and governance. We organised these themes under five broad processes relating to the projects' initiation, organisation, implementation, governance and broader embedding. One can find loose similarities of these processes with those employed in standard project management theory and practice (Koskela and Howell 2002; PMI, n.d.).[1] For the purpose of this article they make sense as they provide useful analytical lenses and structural signposts. Theoretically, they are helpful as they operationalise the "project" addition or qualifier in Roy's and Ong's notion of worlding.

Structured under the five process-dimensions we unfold our analysis in two corresponding narratives featuring China's and India's urbanisation efforts. Both, the five processes and their distinctive features help to show how China and India use the Smart Cities Mission and the Eco-City Initiative to operationalise their "worlding" that reaches from basic local development to international outreach.

2. Initiation and focus

In the following section we will provide a brief historical contextualisation of India's and China's urbanisation efforts to show under what characteristics and circumstances the focus has shifted to cities. This embeds both country's most recent urban initiatives.

2.1. From India's villages to her "urban awakening"[2] in smart cities

To Mahatma Gandhi the cradle of India's awakening was not to be found in the urban space, but was located in her villages. It was here, in the villages of India's vast rural lands where her future and her development emerged. Although Gandhi recognised the importance of urban-industrial economy, his main focus, expressed in his famous Hind Swaraj and Indian Home Rule publications, remained village development (Spodek 1975). For long, Gandhi's legacy on the role of villages dominated India's political and economic decision-making.

India's decision to liberalise its economy during the 1990s and to establish a market – instead of state-centric economy, was not only underpinned by processes of decentralisation and globalisation but also an increased emphasis on the role of the urban sector in development (Ruparelia, Reddy, and Harriss 2011; Hoelscher 2016). Worldwide, cities were increasingly seen as potent sites for "harnessing urbanization for growth and poverty alleviation" – as the title of an important World Bank report phrased it (The Word Bank 2009). Following this thinking the proclamation of "India's urban awakening" did not take long: India's urbanisation potentials were analysed and mapped in a report by McKinsey Global Institute a year later (McKinsey Global Institute 2010). The report underpinned a "market-led approach" to urban development in which efforts in urban transport and infrastructure sectors become key drivers for economic growth (Hoelscher 2016).

The changing focus from villages to cities provides the backdrop for urban development efforts in post-liberalised India. Cities, not villages are seen as the sites in which "development" and "modernisation" are not only kindled but also marketed. Smart cities form the most recent facet of urban development efforts under this approach. According to Hoelscher (2016) the term "smart city" emerged in India in the late 2000s and coincided with "the linking of the private sector and ICT-related [urban] e-governance [as well as the] permeation of the smart city concept in Europe" (32). Similar to Europe, in India, too, the term was soon linked to major transport and infrastructure developments, such as the Industrial Corridors between major Indian cities or the greenfield city projects, of which Dholera or GIFT city are often cited – and contested examples (Datta 2015b). These smart city efforts had been pursued at State level, with Gujarat, Tamil Nadu and Maharashtra as forerunners of States integrating market-led approaches to urban development (Datta 2015b; Hoelscher 2016).

However, it was the in May 2014 newly elected prime minister Narendra Modi of the Bharatya Janata Party (BJP), who tied "smart cities" squarely to India's urban policy agenda at Central level. In June 2015 the new prime minister announced three programmes to tackle urbanisation: "Housing for All",[3] AMRUT,[4] and the "Smart Cities Mission". While Housing for all and AMRUT replace earlier policy programmes, the Smart Cities Mission is a new scheme initiated under the

Modi administration and considered as the new flagship amongst the urbanisation programmes. Under the Smart Cities Mission, which runs from 2015 until 2020, 100 of India's cities are to become smart. With this the Government of India (GOI) aims to tap into the potential of cities as "engines of growth for the economy", expecting that by 2030 40% of India's population will live in urban areas contributing with 75% to the nation's GDP (Ministry of Urban Development – GOI 2015, 5).

2.2. From the countryside to China's eco-city development

China's interest in sustainable and environmentally conscious development is testified since the early 1970s, when the hunger crisis following the Great Leap Forward revealed the limits of its soil (Shapiro 2016). Green urbanisation became a distinct pillar of environmental policy as early as 1973. Institutionally, the overall environmental policy focus was accompanied by the establishment of a national Environmental Protection Office, which was later restructured as the State Environmental Protection Agency and since 2008 acting as the Ministry of Environmental Protection (Baker 2016).

Promoting urbanisation has not always been a priority for Modern China's politics: On the contrary, during the early republic and the Mao era (1949–1978), urbanities were strategically dismantled to break with Ancient traditions, and city people were forcefully relocated to the countryside. This changed poignantly with the launch of a diverse market reforms under Deng Xiaoping, which also encouraged massive urbanisation. By the early 2000s, city development experienced a renaissance within China's overall development efforts. Since then, citizens are politically encouraged to move towards the metropolitan areas along the Yellow Sea coast, especially those of Hebei and Zhejiang province (Liu 2015).

Initiatives from various Chinese government authorities sparked an evolution of city concepts and urbanisation approaches over the last decades, among them smart city initiatives (Yu and Xu 2018). Liu et al. (2014) find that during the 1990s, approaches increasingly foregrounded urban design with "environmental aspects", such as garden city or green city. From the 2000s onwards, more comprehensive umbrella schemes covering several aspects of sustainable development integrated all of these approaches. One prominent concept among them is that of eco-city development, initiated by the Ministry of Environmental Protection. It is the broadest of several Chinese government initiatives to promote eco-city development, addressing a large set of sustainability issues. Two other, yet less comprehensive and smaller approaches are the "low carbon provinces and city program" of the National Development and Reform Committee and the "low carbon eco-city program" of the Ministry of Housing and Urban-Rural Development (De Jong et al. 2016).

3. Organisation

This section will detail specific features that characterise the organisation of India's and China's urban initiatives and show how in both countries political decentralisation plays a central role for organising their urban projects.

3.1. India: competing for becoming smart

One of the core features of the Smart Cities Mission is its competitive approach. In 2015, 100 cities from all over India were shortlisted to take part in a challenge to become a "smart city" under the Smart City Mission policy framework. Municipal officials of these 100 cities were invited to develop smart city plans for their cities and compete with each other to be elected for funding. The competition procedure foresaw several consecutive awarding rounds between 2016 and 2018. Cities with successful proposals could start implementing their plans, while others needed to improve their plans and re-submit for the subsequent of a total of four selection rounds. As highlighted by India's Minister of Urban Development Mr Naidu the Smart Cities Mission "is a landmark in the annals of urban development [in India as] for the first time in the country and perhaps the world, investment in urban development is being made based on the basis of competition" (Naidu 2016).

The competition mechanism on which the Mission is based is noteworthy as it relates to one of Modi's electoral promises to "usher in a new era for Indian federalism" (Arora 2015). The Smart Cities Mission can be seen as Modi's attempt to strengthen the States' role as States and Urban Local Bodies are responsible for developing the Smart Cities Plans according to their own needs and preferences. For the Modi administration this is a move away from the "one-fits-all" approach which had dominated centrally sponsored schemes and which did often not provide a well-suited match between the States' needs and the centrally designed scheme (Sinha 2015).

The devolution of power from India's central level of government to the states has been a lingering issue on India's political agenda (Ruparelia 2015; Sengupta 2015). One of Modi's focal points during his election campaign was to address this and many find that Modi's emphasis on reforming India's governance system along with promoting economic development played a key role for his striking electoral victory at India's central government level in 2014 (Ruparelia 2015).

3.2. China's All-in-One approach for eco-cities

With the election of Xi Jinping as the seventh president of the People's Republic of China, policy efforts soon concentrated on urban development. In one year after Xi took office all eco-urban development became subsumed under the so-called "All-in-One' Pilot Cities" project. This approach is designed as a top-down response to problematic rapid urbanisation. It encompasses a variety of eco-city development projects and serves as an umbrella to green urbanisation. Therein, sustainable (technological) solutions are strongly encouraged and sought for, and corporate investment supported. The action is led by a consortium of National Development and Reform Commission, Ministry of Land Resources, Ministry of Environmental Protection, and Ministry of Housing and Urban-Rural Development (NDRC 2014). It suggests to embrace a comprehensive approach for enforcing regulations, promoting energy-saving and structural adjustments within a frame of poverty alleviation. Encompassing 28 city and county level eco-initiatives, the "All-in-One" project recommends diversity in solving urban environmental problems. One of its core characteristics is the absence of a blueprint model for city development.

This corresponds with an overall trend towards decentralisation (Rong, Jin, and Long 2015). Following the previous Hu regime's directions, Xi continues devolving power to local authorities (Baker 2016). Transferring urban development responsibilities to municipalities suggests these measures are linked to bottom-up reforms. Therefore, green urbanisation shapes political restructuring and adaptation (Sharifi 2016). Decentralisation efforts are interconnected with green economic development, especially through resource efficiency (Baker 2016). Some members of president Xi's cabinet vigorously promote green urbanisation within the realms of main economic priorities (PRCNC 2016). Eco-cities have thus become an integral part of the economic change that aims at replacing the comparatively young, yet carbon-heavy technologies on which the domestic economy relies, with green alternatives (MEP 2016). Promoting localised green economies for eco-city development are characteristic facets of the "All-in-One" approach.

One key organising mechanism in China's eco-city development is the use of indicators to benchmark and asses performance. The Ministry of Environmental Protection was one of the first to develop specific indicators in support of their eco-city development initiative that broadly cover economic and social development as well as environmental protection (De Jong et al. 2016). Based on this, the Ministry accredited hundreds of eco-cities throughout the country, yet there are many initiatives that are being developed by local authorities without official recognition by China's central authorities (De Jong et al. 2016).

4. Implementation

In this section we will show which tensions arise and are incorporated in the implementation of India's and China's urban projects.

4.1. Indian smart cities: from basic infrastructure to smart technologies

The Smart Cities Mission document refrains from defining a smart city as "there is no universally accepted definition [and] even [within] India there is no one way of defining a Smart City" (Ministry of Urban Development – GOI 2015, 5). Despite this definitional openness, one of the dominant facets of "smartness" under the Smart Cities Mission is that it is to be operationalised within the context of basic infrastructure development.

India's cities, both small and large, struggle with deficits in core infrastructure and service provision, which are, however, significantly more pronounced in small-sized cities with a population of 1 million inhabitants or less (Nandi and Gamkhar 2013). By targeting 100 cities the Smart Cities Mission does not bring India's well-known urban metropolises of Delhi, Kolkatta or Mumbai, but smaller cities such as Shimoga, Ludhiana or Guwahati and their infrastructural shortcomings into focus. Adequate water supply, assured electricity supply, sanitation and public transport are considered elements of core infrastructure a smart city in India should provide. According to the GOI, such urban infrastructure development can be realised incrementally, by "adding on layers of smartness" (Ministry of Urban Development – GOI 2015, 5).

Smartness is envisioned to be built up in three layers that build on each other (Sharma 2016): after the provision of city-wide core infrastructure, smart solutions are to be applied to improve infrastructure and services, chiefly by adding ICT-based applications (e.g. application of smart meters). Besides city-wide infrastructure development and their ICT-based upgrades, efforts to "smarten" the city also include area-based approaches, which concentrate efforts in certain areas instead of implementing isolated projects throughout the city (Aijaz and Hoelscher 2015). Though retro-fitting (improving existing), re-development (replacing existing) or greenfield development (building new areas) these areas are to become lighthouses and function as models for replication for other areas within the city but also beyond.

Hoelscher (2016) argues that the formulation of the Smart Cities Mission has undergone a "shift in discourse from smart cities as [Special Economic Zones][5] that would attract global investment and house the urban elite, to smart cities being more actively framed as inclusive projects within existing cities" (34). Initially, Modi's narrative when he was still campaigning for becoming prime minister, stressed greenfield developments through building 100 *new* cities. The "Right to Fair Compensation and Transparency in Land Acquisition, Rehabilitation and Resettlement (LARR) Act", played an important role in this focus-shift. The Act was issued under the former government and strengthened the rights of the landowners, by requiring consultation procedures of village governments and land owners and stipulating certain compensation and rehabilitation measures (Government of India 2013). Modi's administration did not succeed in relaxing the Act's consultation and impact procedures (Hoelscher 2016). As a consequence the Smart City Mission had to be reformulated by including approaches targeting existing (parts of the) cities.

With this, the Smart City Mission incorporates the contested focus of urban development efforts and the resulting tension between providing basic infrastructure and realising smart, i.e. ICT-based, solutions. Resorting to the idea of incrementalism (by adding layers of smartness) is the Mission's approach to bridge this. Different to smart city developments elsewhere, smartness under the Smart Cities Mission is less related to connotations of green and sustainable urban development. Instead, improving effectiveness and efficiency through IT connectivity and digitisation come to the fore. Not the routine basic infrastructure efforts, but ICT-based solutions and projects implemented in either green- or brownfield development-sites are seen as the pinnacles of the Smart City Mission, as they can function as showcases for and lighthouses of smart urban developments.

4.2. Chinese eco-cities: between public propaganda and failure of delivery

Encouraged and promoted by policy attention, targets and funding, hundreds of eco-city sites have mushroomed across the country (Yu 2014). They are embedded in a public discourse fostering a variety of developments ranging from an environmental protection law to funding eco-friendly start-ups and technology. Eco-city projects experiment with progressive urban planning and transportation infrastructure, even comply with green architectural standards and alternative energy concepts or smart technology (Hoffmann 2011). In this public discourse eco-cities test sustainable technology and other measures and contribute to redefine Chinese national identity while demarcating itself from others (Wong 2011).

One of the eco-city flagships is Tianjin Eco-city in Hebei province. Starting 2008, Tianjin is a joint-project by the governments of Singapore and China. The city stands on saline-alkaline land and therefore exemplifies both the necessity and possibility to build green cities even on non-arable and water-scarce land (Yu 2014). Through technologically advanced means of urban planning, the Party aims at cleaning up previous environmental damages (PRCNC 2016). Despite positive official propaganda and support, Tianjin Eco-City struggles to fulfil its ecological promises and is not accepted by the targeted young middle-class urbanites (Kaiman 2014). It remains a high-profile ghost town close to Beijing up until today and is not the only negative example (Caprotti, Springer, and Harmer 2015).

The discrepancy of vision and material reality is captured in May Hald's book that provides a sobering view of eco-city development at a location close to Shanghai: "There was nothing even alluding to an eco-city construction at Dongtan site. None of the residents I spoke with on Chongming Island had heard of the eco-city project" (2009, 11). Despite government propaganda and support for green urban development, three main strands of critique on China's eco-cities manifested themselves: misappropriation of funds, detrimental environmental effects and problems of privatisation.

Van Rooij (2006) and Chien (2013) point to the problem of central government funding being soaked up by mid-level provincial and municipal governments. Research on the state and development of Chinese eco-cities describes the effects of illegal administrative practices in distributing public and private funds, and estimates the damage for both governance and public health (Hald 2009). Additionally, many of the eco-city projects struggle with severe problems deriving from their immediate natural environment. Wong (2011) describes the vulnerable ecosystems of eco-cities, as many are built on once-polluted or non-arable land; some even on reclaimed coastal land or artificial coast strips. These sites are literally threatened by underlying environmental hazards such as soil contamination or sea level rise. As elsewhere, China's eco-city development implicates the governance of real estate and the involvement of private corporations. Chinese local authorities can boost their economies by leasing land to private city developers (De Jong et al. 2016). Large corporates involved in developing eco-cities opened room for real-estate speculation (Van Rooij 2006), often accompanied by an aggressive marketing strategy (Joss and Molella 2013). Such speculation arising with what Goldman frames "speculative urbanism" (Goldman 2011) is one of the central objects of debate. Elsewhere, this phenomenon has been reported as well, leading scholars to expressing alarming concerns about lack of social justice (Datta 2015b).

China's eco-city development highlights tensions between official propaganda and perceived failures in actual implementation. The unbroken official support can be interpreted as a cultural trait of avoiding loss of face through admitting failures, especially on international stage. Domestically, such behaviour creates space for further experimentation as it distracts the focus of press and critics.

5. Governance

This section details the characteristics of governing India's and China's urban initiatives and shows which actors and how they are encouraged to participate.

5.1. India's corporate governance of smart cities

The Smart City Mission stipulates that cities implement their smart city plans under the governance of a Special Purpose Vehicle (SPV) that will "plan, appraise, approve, release funds, implement, manage, operate, monitor and evaluate the Smart City development projects" (Ministry of Urban Development

– GOI 2015, 12). The installation of the SPV must be understood against the backdrop of what some call the "decentralisation deficit" in Indian urban governance (Sivaramakrishnan 2007).

Three layers of governance shape Indian governance of the urban sector. The national level, or Centre, has a facilitative role and issues main policies and directions. Federal states are the key responsible actors for urban governance and it is often at this state level where the responsibility for providing basic amenities and services is located. Local urban governments, called "Urban Local Bodies" (ULBs), only have a restricted role and depend heavily on powers and funds devolved from the states (Perret et al. 2014). Though states are responsible they often cannot effectively deliver services at local level. Besides this, the lack of decentralisation by not granting more financial and operational autonomy to the ULB's contributes to performance and accountability deficits in urban governance (The World Bank 2011).

Though the 74th Constitutional Amendment Act of 1993 has put decentralisation in the urban sector on India's political agenda, progress is slow and has mainly concentrated on devolving administrative functions to the ULBs, keeping them financially dependent (Nandi and Gamkhar 2013; Perret et al. 2014). Institutional inefficiencies aggravate the lack of decentralisation efforts. "A multitude of administrative bodies and bureaucratic rigidities" contribute to making Indian urban governance complex and inert (Nandi and Gamkhar 2013).

The establishment of a SVP as the governance body to implement the smart city plans is a way to bypass these different institutional and administrative layers and inefficiencies. With the establishment of a SVP direct operational and financial responsibilities are installed at the local level allowing room for manoeuvre and discretion. This is done by establishing a structure with "corporate" governance features. The SVPs that are being formed for the winning cities of the smart city competition are "headed by a full time CEO", a board of directors and are established in the form of a "Limited Company" under the Companies Act.

The influence of corporate thinking in urban India must be seen against the backdrop of broader patterns of urban development. Since the 1990s along with liberalising her economy the Government of India started to "increase systematically foreign direct investment (FDI) as percentage of the [gross domestic product]" (Ghertner 2014, 1558). This resulted in a massive increase in investments into urban real estate and infrastructure, bringing about a "new planning regime" shifting from "state-run infrastructure to infrastructure that is run and managed by private developers" (Roy 2009, 77).

Pivotal in these liberalised economic dynamics is the acquisition of land. Municipal authorities have been encouraged to liberalise their land markets to facilitate the investment in and development of infrastructure and real-estate projects (Ghertner 2014). Most of these developments are taking place at the outskirts of the cities, where inflows of capital investment contribute to the urbanisation of peri-urban, rural and protected land, "previously used for commoning, subsistence or other purposes not defined primarily by ground-rent maximization" (Ghertner 2014, 1560).

Despite strong corporate influences, enabled through economic liberalisation, the Smart Cities Mission needed to opening up the focus from exclusive greenfield developments to developing existing urban areas. This was accompanied by the move towards "more inclusive language" (Hoelscher 2016, 35). Critiques pointed to the elitist accents of the envisioned smart cities which were mainly in line with aspirations of the urban middle-class. As a consequence the smart cities rhetoric changed and incorporated the importance of citizen participation in decision-making about urban developments: Consultation of citizens forms one of several assessment criteria of the smart city plan that every city participating in the smart city competition has to submit. Besides participating to a certain degree in the formulation of the smart city plans, the Smart Cities Mission states that "the participation of smart people will be enabled by the SPV through increasing use of ICT, especially mobile-based tools" (Ministry of Urban Development – GOI 2015, 18). These elaborations are, however, very broad and unspecific and stand in stark contrast to the level of detail the Mission stipulates for both the type of smart infrastructure solutions to be applied and the way of organising financial and operational governance through SPVs.

5.2. A broader set of stakeholders and social media for China's eco-city development

The promotion of localised approaches for green urban development showcases how Chinese authorities increasingly allow a broader set of stakeholders to engage in their striving towards sustainable development. In recent years China has acknowledged that promoting green development requires a broad engagement of stakeholders, including civil society. And despite keeping a strong hand, with Xi a relaxation of regulations on civic organisations has occurred (Baker 2016). Utilising the internet and social media, such non-governmental organisations publicise environmental information and offer discussion forums (Baker 2016). Social digital media played a crucial role in building up the eco-cities' bad reputation despite the authorities' positive online and offline propaganda. Digital news about eco-city projects, for example, appear on a regular basis. They evoke images of green and liveable environments and form a stark contrast to smog-infested and toxically contaminated urban realities. Despite their supervision, the internet and social media form the central stage for discussing modern urban imaginaries (Caprotti, Springer, and Harmer 2015). Here, "netizens", representing a societal desire of individuality gained through social attention, media presence, and visibility, shape the public opinion. The public debate around the Tianjin Eco-city stands exemplary for the power of social media platforms like Sina Weibo or Tencent's WeChat in expressing discontent and questioning urban development. The critical online discourse reveals these tensions between imaginary and rhetoric as compared to material delivery and eco-promises. It also exposes the struggles between top-down industrialisation and urbanisation on the one hand and community goals on the other.

Besides carefully opening up highly controlled opportunities for public discussion and engagement, the authorities embrace digital communication as a way to enrol people in their green course. China fosters awareness for social responsibility and the importance to adopt green lifestyles predominantly through digital outlets like subway monitors to TV adverts in addition to traditional placards (PRCNC 2016). Both government and netizens employ social media to raising awareness for eco-civilisation, environmental protection, and advocate cultivating green (urban) lifestyles. With new apps, social credit points, and material incentives, government and also private initiatives target individual practices, for example by involving citizens as reporters of environmental pollution (i.e. MEP 2016). These range from encouraging basic habits like not littering to strategic consumer choices as to invest in energy-efficient, smart technology.

6. Embedding

This section, finally, embeds the Indian Smart City Mission and China's "All-in-One" project within broader political dynamics and aspirations.

6.1. Indian smart cities to "transform-nation"

The Smart Cities Mission is implemented in a landscape of other initiatives and efforts in urban development, such as the earlier mentioned Housing for All or AMRUT scheme, but also programmes for fostering entrepreneurship through skills development among the urban poor or initiatives to counter open defaecation and improve solid waste treatment. The Smart Cities Mission emphasises this convergence with other urban schemes as a way to show the embeddedness and complementarity of smart cities efforts with other economic and infrastructure initiatives (Ministry of Urban Development – GOI 2015).

Yet, the Mission can also be seen as part of a broader, overarching narrative. The symbol of India's Smart Cities Mission is a butterfly whose main contours are pictured through "digital-looking" small grey rectangles dotted on a white background. The centre of the two wings and the two tips of the antennas have the green and the orange colour of the Indian flag. The butterfly symbol is subtitled with the lines "Smart City – MISSION TRANSFORM-NATION" (Ministry of Urban Development – GOI, n.d.). The butterfly stands as a metaphor for the transformation from a crawling caterpillar to a flying butterfly. The sub-title makes this dynamic for India explicit by combining the words "transformation" and "nation" into one hyphenated composite: the nation is transformed through its smart cities.

The promise of "transforming India" accompanied Modi ever since he started campaigning for becoming India's prime minister. For many, this promise is rooted in the achievements during his tenure as chief minister of the state of Gujarat between 2002 and 2014. During this time he established his Gujarat model of development, emphasising good governance, the importance of an investor-friendly environment and infrastructure development. Though "the primacy of trade and commerce" have a much longer history in this region, Modi could link the economic successes to his style of governance and leadership (Kaur 2015). Following the maxim "minimum government, maximum governance" Modi could establish himself as an action-oriented leader who was able to bring about change by following neoliberals ideas of higher growth rates, capital investment flows and good governance (Kaur 2015). Many voted for Modi in India's general elections in the hope that his Gujarat model of development would not only bear fruits of economic development for the state of Gujarat but would, when replicated elsewhere, also transform the economy of India as a whole.

Along with his victory, Modi reinforced the stance on economic reforms by his party: Since the second decade of India's liberalisation the BJP party had increasingly adapted to the neoliberal discourse. This is a remarkable shift as the BJP had long been extremely critically regarding liberalising India's economy (Ruparelia, Reddy, and Harriss 2011; Kaur 2015). By subscribing to neoliberal ideas the BJP even surpassed the Congress Party – the "original party of economic reforms" – which had facilitated the liberalisation reforms in the 1990s (Kaur 2015). One of the first examples of manifesting the transforming India narrative into India's governance system was the replacement of the Planning Commission, a 64-year-old policy-making body. Shortly after assuming office the new prime minister announced during his first Independence Day speech to replace the Planning Commission with a new institution with the programmatic name: National Institution for Transforming India, or NITI Aayog.[6] NITI Aayog is India's "premier policy think tank, providing both directional and policy inputs" for both Centre and the states (NITI Aayog – GOI, n.d.). Other flagship initiatives for transforming India followed suit. Besides the installation of NITI Aayog and the Smart Cities Mission, Digital India and "Make in India" are two other initiatives that explicitly operationalise Modi's ambition of transforming India. Digital India envisions "to transform India into a digitally empowered society and knowledge economy" (Ministry of Electronics & Information Technology – GOI, n.d.).[7] "Make in India" aims to transform India into a global manufacturing hub by facilitating foreign direct investment.

With this the Smart Cities Mission is part of a whole array of efforts launched under Modi to transform India. Since the beginning, Modi tied his transformation-narrative to a neoliberal prefix and the Smart Cities Mission operationalises these transformative ambitions in the urban sector. And while many dimensions of the Smart Cities Mission can be read as a story of unchecked liberalisation, there are (though still very vague) participatory inroads that point to possibilities for countering the dominance of capital-driven urban development India has experienced since the 1990s. The Mission underpins an understanding that transforming India will be done through its cities and in which ICT-based solutions are the primary driver for change. Thereby the Smart Cities Mission plays into what (Datta 2015a) calls "technocratic patriotism", under which being patriotic is bound to the believe in the power of technology. In the Smart City Mission different Indian aspirations coalesce, involving the transformation of development efforts and their governance – and thus – ultimately – the transformation of the nation. What prefixes will be added, removed or enhanced in this story of transformation remains to be seen.

6.2. China's eco-city development for ecological civilisation

When President Xi Jinping took office in 2013, the nation committed itself to building an "ecological civilisation" in response to several decades of pollution and deconstruction caused by rapid and often forceful change. Scholars have increasingly targeted the question of how to revoke the environmental debts acquired during its rapid industrialisation and urbanisation since opening up since 2003 (i.e. Zhu 2004; Muscolino 2009; Li and Liu 2011). By putting ecological civilisation on the agenda, Xi revived a concept that had been lingering in Chinese discourse before. The overarching idea is not only as old as the 1990s (Shen 1994), it picks up a 5000-year historical tradition of civilisation excellence and urban role models. A diplomatic perspective on China's contemporary environmental history attests the 1990s an established national strategy for sustainable development, following the 1980s as a period of basic environmental protection. In the twenty-first century, eco-civilisation offers a long-term strategy vital to further modernisation.

The concept of ecological civilisation is defined to address all issues of the environment comprehensively: industry, traffic, residence, and even the "pattern of society" (Ma 2009). Showcasing China's green motivation, it further tributes to a "scientific outlook on development" that is motivated to par with Western ecological informed sciences (Muscolino 2009) and makes fit for the twenty-first century as "people-centered, fully coordinated, and environmentally sustainable" (UNEP 2016, 3).

Green urban development has become an integral pillar of realising China's ecological civilisation in this strive for modernisation. Eco-cities, thus, feature as a socialist commitment to innovative, coordinated, green, open and shared development (i.e. Liu 2015). The latest economic and social development plan for the Peoples' Republic focusses on urban development, aiming to "develop harmonious and pleasant cities" (PRCNC 2016). Other target areas include gentrification and sanitisation, indicating that growing cities need to address social questions to adhere to sustainable and just standards of urbanisation and development (Moore 2016). President Xi began driving "people-centered green" projects forward right after his election and became the figurehead in promoting green, sustainable urban China.

With Xi, eco-city projects have become a core part of environmental politics which forms one pillar to develop the country's overall ecological civilisation. He revived and underlined the importance of ecological urban development at a time when ecological urban development had come under severe pressure by not delivering on its green promises. By linking eco-city development with ecological civilisation, eco-urban development has become a tangible action to enact and materialise the ambition of creating an ecological civilisation.

7. Conclusion

In this article we have studied India's and China's recent urbanisation efforts in two corresponding narratives. Our aim was not to single out one paradigmatic approach and see how the other matches (or fails to match) this benchmark or role model. Often written in a "rhetoric of superlatives" (McFarlane 2010) these accounts have been criticised for their tendency to reduce findings "to a perfunctory and unenlightening assessment how the others compare to the paradigmatic city" (Beauregard 2003, 190). Rather, through the lens of worlding we identified what Ong describes as "some vision of the world in formation" (Ong 2011, 11). In this concluding section we will summarise and highlight certain findings as they relate and inform the ongoing debate on eco and smart cities in a context of worlding.

In doing so we are informed by McFarlane's (2010) idea of indirect learning. The author cautions against a tendency that reduces learning to an idea of "direct transfer", in which the sole focus is on whether and how it might be useful and applied directly elsewhere (McFarlane 2010). Instead, careful attention to differences can broaden and deepen the spectrum of learning without reverting to pre-defined notions of direct usefulness and transferability. Five analytical dimensions relating to focus, organisation, implementation, governance and embedding have helped us to identify characterising features that describe India's and China's worlding projects in detail. The ensuing section will synthesise these findings. It will show how – despite their differences – both India's and China's urban efforts are firmly anchored in a broader agenda of change that is set out to transform the nation and extend into time.

Both Modi and Xi made urban development one of their political foci when they took office. While in China major urban development efforts are linked with ideas of ecological development, in India urbanisation efforts were soon linked to ideas of smartness and new (Information and Commmunication) technologies. Xi's support for eco-city development came at a moment when eco-cities were surrounded by an increasingly bad reputation as they did not seem to deliver on their promises. Xi gave a new and fresh impetus to China's ecological urbanisation agenda. A similar dynamic can be seen in India. Linking urbanisation with smartness provided a new flavour and thereby a connotation of a new beginning to the Indian urban sector, which has been struggling for decades with performance deficits and underfunding. It is also in line with Modi's political agenda in which he aims to establish a modern ICT-based image of his leadership.

India organised her smart urbanisation on the basis of a competitive model in form of the Smart City Challenge, in which 100 cities "compete to become smart". This model has parallels with the so-called "100 Resilience City" Challenge, an initiative pioneered by the Rockefeller Foundation (Rockefeller Foundation, n.d.), conducted just one year earlier than Modi's city competition. The similarities can be seen as an example for what Ong terms the "worlding practice of modelling" (Ong 2011). In this case however, no "urban model" (such as eco-city) with "established values understood as desirable and achievable" but a model to organise urbanisation is circulated globally (Ong 2011, 14). In China, the trend towards decentralisation is mirrored in the country's "All-in-One" approach, which does not foresee one blueprint model for urban development. Under the banner of eco-city hundreds of variants of eco-cities have and are being developed to hatch successful models for replication. Their indexing system can be seen as a "modelling technology" for "standard-setting forms and norms" (Ong 2011, 15) that help the implementation of a variety of successful urban innovations and lifestyles elsewhere.

In the implementation of India's urbanisation efforts, the lure of ICT solutions seems to compete with the mundane necessity to provide basic infrastructure, especially for the masses of India's poor (er) urban dwellers. The attractiveness of smart ICT can be seen in its promise as the central driver or even "leap-frogger" for change. And while elsewhere "smart" is quickly linked to connotations of environmental friendliness and forms of greenness, in India smart technologies are first and foremost seen as change-agents for digital connectivity. China's eco-city Tianjin stands as an example for the challenges that emerge when implementing eco-urban ambitions (Caprotti 2014). It exemplifies the tension between public promises and failure of delivery.

The trend of decentralisation in both China's and India's governance arenas suggests a move towards a broader participation of stakeholders. Yet, in the urban context the main addresses of this seem to have been corporate stakeholders. In India this has resulted in the devolution of power to public–private partnerships that govern urbanisation projects. Both in China and in India this put the role of civil society into the centre of attention – and tension. It is remarkable how despite close control by Chinese authorities netizens have emerged as a force that can shape the public discourse on urbanism.

With our analysis we have not only detailed a spectrum of features that characterises China's and India's focus, organisation, implementation and governance of their respective urban efforts. Importantly, the analysis of India's Smart City Mission and China's All-in-One project has shown how both efforts are embedded in broader agendas of change. The Indian Smart City Mission is part of a broader ambition to transform the nation enabling her "smart incarnation" in modernity. The aspiration of reaching modernity is to be realised through embracing smartness manifested in ICT and new technologies, especially as these technologies hold the promise of leapfrogging into the modern state. India's firm belief in the power of technology is also fuelled by her remarkable success as a global IT power house through and with which it entered the world's market stage in the twenty-first century.

China's approach to urban development is integrated within the broader framework of ecological civilisation. Xi revived the framework by explicitly linking eco-urbanism to it as a means to enact ecological civilisation on the (urban) ground. The term "ecological" instead of "environmental" indicates

China's attempt to distinguish itself from other countries in a global race to becoming the greenest nation. In this context, the term "ecological" allows to connect philosophically with traditional Confucian and Taoist values. The former build on rigid rules and emphasise social order, while the latter rely on harmony between human will and nature (Shen 1994). Additionally, an "ecological" framing grants a fresh start into environmental governance. This new terminology tributes to scientific principles and solutions that Beijing utilises in a historicist sense to reconstruct and rejuvenate the nation state's twenty-first-century identity that is torn between communist past and capitalist present.

With India embracing the idea of transformation with smart technologies as key drivers of change and China subscribing to the realisation of an ecological civilisation their cities will be one of the central sites where these aspirations are being enacted and where these same aspirations will be confronted with other (urban) ideas of becoming.

Notes

1. These typically relate to initiating, planning, executing, controlling, and closing.
2. "India's urban awakening" is the title of an influential McKinsey report on Indian urbanisation (McKinsey Global Institute 2010).
3. "Housing for All" is targeted especially at the weaker section of Indian society, and sets out to construct houses in urban areas for 20 million families for the urban poor in the next 7 years.
4. AMRUT or "Atal Mission for Rejuvenation and Urban Transformation" is an infrastructure scheme aiming to provide basic services (e.g. water supply, sewerage, urban transport) to households and build amenities in cities to improve the quality of life.
5. Special Economic Zones are greenfield development projects and enclaves "with the minimum possible regulations (…) to overcome the shortcomings experienced on account of the multiplicity of controls and clearances" and to attract "world-class infrastructure" enabled by "larger foreign investments in India" (Ministry of Commerce and Industry – GOI, n.d.).
6. The abbreviation NITI is also a word play, as it is the Hindi word for "policy". Aayog is Hindi for "commission".
7. Modi's heavily criticized major policy action of demonetizing the 500 and 1000 Rupee banknotes is in line with this ambition. Announced as an effort to counter the shadow economy this initiative must also be seen as an attempt by the Modi administration to advance the "digitization of India", by transforming her into a digitally based economy.

Disclosure statement

No potential conflict of interest was reported by the authors.

References

Agyeman, J., R. D. Bullard, and B. Evans. 2003. *Just Sustainabilities. Development in an Unequal World*. London: Earthscan.
Aijaz, R., and K. Hoelscher. 2015. "India's Smart Cities Mission: An Assessment." *Observer Research Foundation (ORF) Issue Brief* 124: 1–12.
Arora, B. 2015. "The Distant Goal of Cooperative Federalism." *The Hindu*. Accessed March 21, 2017. http://www.thehindu.com/opinion/op-ed/the-distant-goal-of-cooperative-federalism/article7232184.ece.
Baker, S. 2016. *Sustainable Development*. 2nd ed. New York: Routledge.
Beauregard, R. A. 2003. "City of Superlatives." *City and Community* 2 (3): 183–199.
Brenner, N. 2001. "World City Theory, Globalization and the Comparative-Historical Method-Reflections on Janet Abu-Lughod's Interpretation of Contemporary Urban Restructuring." *Urban Affairs Review* 37 (1): 124–147.
Brenner, N., and R. Keil. 2006. *The Global Cities Reader*. New York, NY: Routledge.
Caprotti, F. 2014. "Eco-urbanism and the Eco-city, or, Denying the Right to the City?" *Antipode* 46 (5): 1285–1303. doi:10.1111/anti.12087.
Caprotti, F., C. Springer, and N. Harmer. 2015. "'Eco' for Whom? Envisioning Eco-urbanism in the Sino-Singapore Tianjin Eco-city, China." *International Journal of Urban and Regional Research* 39 (3): 495–517.
Chien, S. 2013. "Chinese Eco-cities: A Perspective of Land-Speculation-Oriented Local Entrepreneurialism." *China Information* 27 (2): 173–196.

Datta, A. 2015a. "A 100 Smart Cities, a 100 Utopias." *Dialogues in Human Geography* 5 (1): 49–53.

Datta, A. 2015b. "New Urban Utopias of Postcolonial India: 'Entrepreneurial Urbanization' in Dholera Smart City, Gujarat." *Dialogues in Human Geography* 5 (1): 3–22.

De Jong, M., C. Yu, S. Joss, R. Wennersten, L. Yu, X. Zhang, and X. Ma. 2016. "Eco City Development in China: Addressing the Policy Implementation Challenge." *Journal of Cleaner Production* 134 (Part A): 31–41.

Ghertner, D. A. 2014. "India's Urban Revolution: Geographies of Displacement beyond Gentrification." *Environment and Planning A: Economy and Space* 46: 1554–1571.

Goldman, M. 2011. "Speculating on the Next World City." In *Worlding Cities: Asian Experiments and the Art of Being Global*, edited by A. Roy and A. Ong, 229–258. Oxford: Wiley-Blackwell.

Government of India. 2013. *Right to Fair Compensation and Transparency in Land Acquisition, Rehabilitation and Resettlement Act*. New Delhi: Ministry of Law and Justice – GOI.

Hald, M. 2009. "Sustainable Urban Development and the Chinese Eco-city. Concepts, Strategies, Policies and Assessments." Unpublished dissertation. University of Oslo.

Hoelscher, K. 2016. "The Evolution of the Smart Cities Agenda in India." *International Area Studies Review* 19 (1): 28–44.

Hoffmann, L. 2011. "Urban Modeling and Contemporary Technologies of City-Building in China: The Production of Regimes of Green Urbanisms." In *Worlding Cities: Asian Experiments and the Art of Being Global*, edited by A. Roy and A. Ong, 55–76. Oxford: Wiley-Blackwell.

Hollands, R. G. 2008. "Will the Real Smart City Please Stand Up?" *City* 12 (3): 303–320. doi:10.1080/13604810802479126.

Joss, S., and A. Molella. 2013. "The Eco-city as Urban Technology: Perspectives on Caofeidian International Eco-city (China)." *Journal of Urban Technology* 20 (1): 115–137.

Kaiman, J. 2014. "China's 'Eco-cities': Empty of Hospitals, Shopping Centres and People." *theguradian.com*. Accessed February 5, 2016. https://www.theguardian.com/cities/2014/apr/14/china-tianjin-eco-city-empty-hospitals-people.

Kaur, R. 2015. "Good Times, Brought to You by Brand Modi." *Television & New Media* 16 (4): 323–330.

Koskela, L. J., and G. Howell. 2002. "The Underlying Theory of Project Management Is Obsolete." In *Proceedings of the PMI Research Conference*, 293–302. PMI.

Li, X., and Y. Liu. 2011. "The Current Situations, Problems and Solutions of Chinese Eco-cities Development." [中国低碳生态城市发展的现状、问题与对策.] *Urban Planning Forum* [城市规划学刊] 4: 23–29.

Liu, T. 2015. "Case Studies of Eco-cities in Coastal Region of Southeast China." Paper prepared for EAEH. Unpublished Manuscript.

Liu, H., G. Zhou, R. Wennersten, and B. Frostell. 2014. "Analysis of Sustainable Urban Development Approaches in China." *Habitat International* 41: 24–32.

Ma, D. 2009. "Study on the Construction Path and Evaluation System of Eco-civilized City." [生态文明城市构建路径与评价体系研究.] *Urban Studies* [城市发展研究] 10: 80–85.

Marvin, S., A. Luque-Ayala, and C. McFarlane. 2016. *Smart Urbanism: Utopian Vision or False Dawn?* New York: Routledge. doi:10.1017/CBO9781107415324.004.

McCann, E., A. Roy, and K. Ward. 2013. "Urban Pulse-Assembling/Worlding Cities." *Urban Geography* 34 (5): 581–589. doi:10.1080/02723638.2013.793905.

McFarlane, C. 2010. "The Comparative City: Knowledge, Learning, Urbanism." *International Journal of Urban and Regional Research* 34 (4): 725–742.

McKinsey Global Institute. 2010. "India's Urban Awakening: Building Inclusive Cities, Sustaining Economic Growth." Mumbai, London, Soul, San Francisco.

(MEP) Ministry of Environmental Protection The People's Republic of China. 2016. "Mission." Accessed March 27, 2017. http://english.sepa.gov.cn/About_SEPA/Mission/.

Ministry of Commerce and Industry – GOI. n.d. "Special Economic Zones – Introduction." Accessed March 26, 2017. http://www.sezindia.nic.in/about-introduction.asp.

Ministry of Electronics & Information Technology – GOI. n.d. "Digital India – Introduction." Accessed March 26, 2017. http://www.digitalindia.gov.in/content/introduction.

Ministry of Urban Development – GOI. 2015. "Mission Statement & Guidelines for Smart Cities." http://smartcities.gov.in/writereaddata/SmartCityGuidelines.pdf.

Ministry of Urban Development – GOI. n.d. "The Smart Cities Mission." Accessed March 26, 2017. http://smartcities.gov.in/content/.

Moore, S. A. 2016. "Testing a Mature Hypothesis: Reflection on 'Green Cities, Growing Cities, Just Cities: Urban Planning and the Contradiction of Sustainable Development'." *Journal of the American Planning Association* 82 (4): 385–388.

Muscolino, M. 2009. "Global Dimensions of Modern China's Environmental History." *World History Connected* 6 (1), Accessed February 10, 2017. http://worldhistoryconnected.press.illinois.edu/6.1/muscolino.html/.

Naidu, V. 2016. "A Mission to Transform the Nation through Cities." *The New Indian Express*. Accessed February 10, 2017. http://www.newindianexpress.com/opinions/2016/feb/01/A-Mission-to-Transform-the-Nation-through-Cities-886970.html.

Nandi, S., and S. Gamkhar. 2013. "Urban Challenges in India: A Review of Recent Policy Measures." *Habitat International* 39: 55–61.

NDRC (The People's Republic of China National Development and Reform Commission). 2014. "Notice on Implementing the Plans Integration Pilot Work in Cities and Counties."

NITI Aayog – GOI. n.d. "NITI Aayog – Overview." Accessed March 26, 2017. http://niti.gov.in/content/overview.

Ong, A. 2011. "Introduction: The Art of Being Global." In *Worlding Cities: Asian Experiments and the Art of Being Global*, edited by A. Roy and A. Ong, 1–26. Oxford: Wiley-Blackwell.

Ong, A., and S. Collier. 2005. *Global Assemblages. Technology, Politics and Ethics as Anthropological Problems*. Oxford: Blackwell.

Perret, L., Ravikant Joshi, Abhay Kantak, Ravi Poddar, Achin Biyani, Ram J. Khandelwal, and Romain Fayoux. 2014. "Panorama of the Urban and Municipal Sector in India." New Delhi.

Pickvance, C. 1986. "Comparative Urban Analysis and Assumptions About Causality." *International Journal of Urban and Regional Research* 10: 162–184.

PMI (Project Management Institute). n.d. Accessed May 14, 2019. https://www.pmi.org/pmbok-guide-standards/foundational/pmbok.

PRCNC (The People's Republic of China National Congress). 2016. "The 13the Five-Year-Plan for Economic and Social Development of the People's Republic of China (2016–2020)."

Robinson, J. 2011. "Cities in a World of Cities: The Comparative Gesture." *International Journal of Urban and Regional Research* 35 (1): 1–23.

Rockefeller Foundation. n.d. "100 Resilient Cities." Accessed March 22, 2017. http://www.100resilientcities.org.

Rong, X., Y. Jin, and Y. Long. 2015. "Understanding Beijing's Urban Land Use Development from 2004–2013 Through Online Administrative Data Sources." In *Recent Developments in Chinese Urban Planning : Selected Papers From the 8th International Association for China Planning Conference*, Guangzhou, China, June 21–22, 2014, edited by Q. Pan and J. Cao, 183–218. Cham: Springer.

Roy, A. 2009. "Why India Cannot Plan Its Cities: Informality, Insurgence and the Idiom of Urbanization." *Planning Theory* 8 (1): 76–87.

Roy, A., and A. Ong, eds. 2011. *Worlding Cities: Asian Experiments and the Art of Being Global*. Oxford: Wiley-Blackwell.

Ruparelia, S. 2015. "'Minimum Government, Maximum Governance': The Restructuring of Power in Modi's India." *South Asia: Journal of South Asian Studies* 38 (4): 755–775.

Ruparelia, S., S. Reddy, and J. Harriss, eds. 2011. *Understanding India's New Political Economy: A Great Transformation?* New York: Routledge.

Sengupta, M. 2015. "Modi Planning: What the NITI Aayog Suggests About the Aspirations and Practices of the Modi Government." *South Asia: Journal of South Asian Studies* 38 (4): 791–806.

Shapiro, J. 2016. *China's Environmental Challenges*. 2nd ed. Polity Press. Hoboken: John Wiley & Sons.

Sharifi, A. 2016. "From Garden City to Eco-urbanism: The Quest for Sustainable Neighborhood Development." *Sustainable Cities and Society* 20. doi:10.1016/j.scs.2015.09.002.

Sharma, S. 2016. "The Idea and Practice of Smart Cities in India." Accessed March 21, 2017. https://blog.mygov.in/editorial/the-idea-and-practice-of-smart-cities-in-india/.

Shen, S. 1994. "Ecological Civilization and Its Theoretical and Practical Basis." [生态文明及其理论与现实基础.] *Journal of Peking University (Humanities and Social Sciences)* [北京大学学报: 哲学社会科学版] 3: 31–37.

Sinha, S. 2015. "Modi Assures Centre Will Move Away From 'One Size Fits All' Schemes." *The Hindu*. Accessed March 21, 2017. http://www.thehindubusinessline.com/economy/states-must-focus-on-growth-investment-efficiency-pm/article6871182.ece.

Sivaramakrishnan, K. C. 2007. "Democracy in Urban India." Accessed March 26, 2017. https://lsecities.net/media/objects/articles/democracy-in-urban-india/en-gb/.

Söderström, O., T. Paasche, and F. Klauser. 2014. "Smart Cities as Corporate Storytelling." *City* 18 (3): 307–320. doi:10.1080/13604813.2014.906716.

Spodek, H. 1975. "From 'Parasitic' to 'Generative': The Transformation of Post-Colonial Cities in India." *Journal of Interdisciplinary History* 5 (3): 413–443.

The Word Bank. 2009. "Systems of Cities Integrating National and Local Policies Connecting Institutions and Infrastructure." Washington, DC.

The World Bank. 2011. "Urbanization in India: Integral Part of Economic Growth." Accessed March 26, 2017. http://web.worldbank.org/archive/website01291/WEB/0__CO-22.HTM.

UNEP. 2016. "Green Is Gold: The Strategy and Actions of China's Ecological Civilization."

United Nations – Department of Economic and Social Affairs. 2016. "The World's Cities in 2016: Data Booklet." Accessed March, 21 2017. https://doi.org/10.18356/8519891f-en.

United Nations – Department of Economic and Social Affairs. 2017. "World Population Prospects: The 2017 Revision." New York.

Van Rooij, B. 2006. *Regulating Land and Pollution in China Lawmaking, Compliance and Enforcement; Theory and Cases*. Leiden: Leiden University Press.

Vanolo, A. 2014. "Smartmentality: The Smart City as Disciplinary Strategy." *Urban Studies* 51 (5): 883–898. doi:10.1177/0042098013494427.

Wong, T.-C. 2011. "Eco-cities in China: Pearls in the Sea of Degrading Urban Environments?" In *Eco-city Planning. Policies, Practice and Design*, edited by T.-C. Wong and B. Yuen, 131–150. Dordrecht: Springer.

Yu, L. 2014. "Low Carbon Eco-city: New Approach for Chinese Urbanisation." *Habitat International* 44: 102–110.

Yu, W., and C. Xu. 2018. "Developing Smart Cities in China: An Empirical Analysis." *International Journal of Public Administration in the Digital Age* 5 (3): 76–91.

Zhu, K. 2004. "On Construction of Eco-civilization." [论建设生态文明.] *China Environmental Protection Industry* [中国环保产业] 8: 6–7.

Index